PERGAMON INTERNATIONAL LIBRARY
of Science, Technology, Engineering and Social Studies

*The 1000-volume original paperback library in aid of education,
industrial training and the enjoyment of leisure*

Publisher: Robert Maxwell, M.C.

TOWARDS GLOBAL ACTION
FOR
APPROPRIATE TECHNOLOGY

THE PERGAMON TEXTBOOK
INSPECTION COPY SERVICE

An inspection copy of any book published in the Pergamon International Library will
gladly be sent to academic staff without obligation for their consideration for course
adoption or recommendation. Copies may be retained for a period of 60 days from
receipt and returned if not suitable. When a particular title is adopted or recommended
for adoption for class use and the recommendation results in a sale of 12 or more copies,
the inspection copy may be retained with our compliments. The Publishers will be
pleased to receive suggestions for revised editions and new titles to be published in this
important International Library.

Other Titles of Interest

BALASSA, B.
Policy Reform in Developing Countries

COLE, S.
Global Models and the International Economic Order

JOLLY, R.
Disarmament and World Development

LASZLO, E.
The Inner Limits of Mankind: Heretical Reflections on Today's Values,
Culture and Politics

LASZLO, E. & BIERMAN, J.
Goals in a Global Community, Volume 1
Goals in a Global Community, Volume 2

MENON, B.P.
Global Dialogue: The New International Economic Order

PECCEI, A.
The Human Quality

TICKELL, C.
Climatic Change and World Affairs

WENK, E.
Margins for Survival: Overcoming Political Limits in Steering Technology

TOWARDS GLOBAL ACTION
FOR
APPROPRIATE TECHNOLOGY

Edited by

A.S. BHALLA

Foreword by

JAN DE KONING

PERGAMON PRESS

OXFORD · NEW YORK · TORONTO · SYDNEY · PARIS · FRANKFURT

U.K.	Pergamon Press Ltd., Headington Hill Hall, Oxford OX3 0BW, England
U.S.A.	Pergamon Press Inc., Maxwell House, Fairview Park, Elmsford, New York 10523, U.S.A.
CANADA	Pergamon of Canada, Suite 104, 150 Consumers Road, Willowdale, Ontario M2J 1P9, Canada
AUSTRALIA	Pergamon Press (Aust.) Pty. Ltd., P.O. Box 544, Potts Point, N.S.W. 2011, Australia
FRANCE	Pergamon Press SARL, 24 rue des Ecoles, 75240 Paris, Cedex 05, France
FEDERAL REPUBLIC OF GERMANY	Pergamon Press GmbH, 6242 Kronberg-Taunus, Pferdstrasse 1, Federal Republic of Germany

First edition 1979

British Library Cataloguing in Publication Data

Expert Meeting on International Action for Appropriate Technology, *Geneva, 1977*
Towards global action for appropriate technology.
- (Pergamon international library).
1. Technology - societies, etc. - Congresses
2. International agencies - Congresses   3. Technology - International cooperation - Congresses
I. Title II. Bhalla, A S
606'.01   T6   78-41191
ISBN 0-08-024305-3 Hard cover
ISBN 0-08-024277-4 Flexi cover

In order to make this volume available as economically and as rapidly as possible the author's typescripts have been reproduced in their original forms. This method unfortunately has its typographical limitations but it is hoped that they in no way distract the reader.

Printed and bound at William Clowes & Sons Limited Beccles and London

CONTENTS

                                                        Page

Foreword by Jan de Koning   xi

Introduction by A.S. Bhalla   xiii

PART I   : CONCEPTS, CRITERIA, STRATEGIES

CHAPTER 1 : APPROPRIATE TECHNOLOGY: SOME CRITERIA   1
             by N. Jéquier

             Introduction   1

             Some Criteria of Appropriateness   4

                Systems Independence of New Technology 4

                Image of Modernity   6

                Individual Technology Versus
                Collective Technology   8

                Cost of Technology   10

                Risk Factor   14

                Evolutionary Capacity of Technology   17

                Single-Purpose and Multi-Purpose
                Technology   20

             Conclusion   22

CHAPTER 2 : TECHNOLOGIES APPROPRIATE FOR A BASIC
             NEEDS STRATEGY   23
             by A.S. Bhalla

             Introduction   23

             A Basic Needs Approach and Strategy   25

             Technological Contents of a 'BN'
             Strategy   29

                Incomes to the Poorest   30

                Access   32

Physical Production of Goods
and Services   36

Participation and Decentralised
Production   39

National and Collective Self-Reliance   42

Technologies Appropriate for Basic
Needs Satisfaction   46

A Suitable Policy Frame   52

National and Local Policies   52

International Policies   57

Concluding Remarks   60

PART II   : EXISTING INSTITUTIONAL FRAMEWORK

CHAPTER 3 : NATIONAL AND REGIONAL TECHNOLOGY GROUPS
            AND INSTITUTIONS: AN ASSESSMENT   63
            by A.K.N. Reddy

Introduction   63

Appropriate Technology Groups and
Institutions   67

Appropriate Agricultural Technology
Cell   69

Cell for the Application of Science
and Technology   70

Appropriate Technology Development
Organisation   72

Council of Scientific and Industrial
Research   73

Development Technology Centre   74

East African Industrial Research
Organisation   75

ESCAP Regional Centre for Technology
Transfer   76

International Rice Research Institute   78

Korean Institute of Science and
Technology   82

Technology Consultancy Centre   84

Towards a Framework of Analysis   86

Development of Appropriate
Technologies   86

Dissemination of Appropriate
Technologies   103

Criteria for Assessment of Groups
and Institutions   109

Assessment of Groups and Institutions   116

Conclusions   127

Appendix: Questionnaire   131

CHAPTER 4 : ACTIVITIES OF THE UN SYSTEM ON
            APPROPRIATE TECHNOLOGY   138
            by W.M. Floor

Introduction   138

Some Existing Coordinating Mechanisms   140

Advisory Committee on the
Application of Science and
Technology to Development   140

Committee on Science and Technology
for Development   141

ACC Sub-Committee on Science and
Technology   142

Review of Ongoing UN Activities   143

UNIDO   145
UNCTAD   146
World Bank   148
ILO   149
FAO   152
UN Economic Commissions   153
UNDP   154
UNEP   155
WHO   156

Better Coordination within the System   159

Concluding Remarks   161

CHAPTER 5 : INTERNATIONAL MECHANISMS FOR
            APPROPRIATE TECHNOLOGY   164
            by F. Stewart

            Introduction   164

            Some Proposals for International
            Mechanisms   171

                Inter-Agency Network for Exchange
                of Technological Information   171

                Industrial and Technological
                Information Bank   173

                Technology Referral Service   175

                Socially Appropriate Technology
                Information System   177

                Transnational Network for
                Appropriate Technologies   177

                World Plan of Action Fund   177

                Sir Austin Robinson's Proposals
                for UNDP   178

                World Employment Conference Proposals
                for New International Mechanisms for
                Appropriate Technology   179

                An International Centre for
                Appropriate Technology   181

                An Interim Global Project towards
                an International Council for
                Appropriate Industrial Technology   183

                World Technological Development
                Authority   183

                USAID Proposal for a Program on
                Appropriate Technology   184

                Proposal of a Working Party of the
                UK Ministry of Overseas Development   186

                Consultative Groups   187

                    Consultative Group on International
                    Agricultural Research   187

                    Special Programme for Research and
                    Training in Tropical Diseases   188

Consultative Group on Food
Production and Investment in
Developing Countries 189

Functions of International Mechanisms 192

    Information 194

    Research and Development 196

    Social/Economic Research 199

Proposed and Existing Mechanisms:
Comparative Analysis 200

    Information 200

    Research and Development 202

    Social/Economic Research 203

Concluding Remarks 204

PART III : GLOBAL ACTION

CHAPTER 6 : A BLUEPRINT FOR ACTION 207
    by P. Henry, A. Reddy, F. Stewart

Introduction 207

Need for Global Action 210

A New International Mechanism for
Appropriate Technology 212

    Functions of the Mechanism 212

    Operation 214

    Organisational Structure 216

        Secretariat 216

        Location 217

        Governing Body 218

        Executive Council 218

    Level of Finance 219

Conclusion 220

FOREWORD

In the past few years many discussions have been held on
the basic needs approach in both rich and poor countries.
A growing number of countries and international organisations have decided to give more emphasis in their aid programmes to assisting the poorest countries and the poorest
groups within countries. Several developing countries have
likewise changed their priorities in order to put more
emphasis on improving the living conditions of their poorest
inhabitants.

This change in priorities has important consequences for the
choice of technologies. The necessity to make use of the
available means of production in such a way that the elementary needs of the population can be satisfied adequately,
presents a considerable challenge to developing countries
and donor agencies. Application of technologies developed
elsewhere will not necessarily lead to the best results and
it may even be counter-productive.

While the need for new and appropriate technologies is
evident, the majority of current research is oriented towards
the type of capital-intensive production found in industrialised countries. An extra effort is necessary to promote and
implement technologies more appropriate to the needs of the
developing countries.

For this reason, the Netherlands Government, in collaboration
with the ILO, convened a meeting in December 1977 at which
individual experts and representatives of developing and developed countries as well as representatives of international
organisations were invited to advise on the possibilities of
organising a better coordinated international effort in the
field of appropriate technology. The group formulated terms
of reference for a team to study the feasibility of establishing a new international mechanism for the promotion of
appropriate technology. The proposals of the feasibility
team were based on extensive consultations in the first half
of 1978 with authorities and experts in developed and de-

veloping countries as well as with the relevant staff of
a selected number of agencies within the UN system.

The revised discussion papers for the meeting and the proposals of the feasibility team presented in this book are
edited by Ajit Bhalla of the ILO. I hope this book will
serve as a basis for thought, and hopefully, for concerted
action.

The Netherlands Government trusts that these proposals will
find widespread support. It is my hope that at the founding
conference recommended by the feasibility team, of which the
Netherlands Government will be one of the sponsors, a new
international mechanism for the promotion of appropriate
technology will be established.

                           Jan de Koning
The Hague,   Minister for Development Cooperation
September 1978   The Netherlands

INTRODUCTION

by

A. S. Bhalla[1]

In recent years the concept of appropriate technology has
gained currency in both developing and developed countries.
This concept has emerged in response to a recognition that,
in spite of rapid rates of economic growth in the past decades, the objectives of employment creation and elimination
of poverty have not been achieved. One of the reasons for
this situation has been an over-emphasis in the past on the
part of a large number of developing countries on capital-
intensive "heavy" industrialisation and the use of techniques that do not necessarily reflect factor endowments
and socio-economic conditions prevailing in these countries.

While the concept of appropriate technology has now come of
age, the national and international action required for its
implementation is not yet commensurate with the magnitude
of the task and the challenge it poses. As will be clear
from Part II of this volume, while a lot of concern has
been expressed to do something about appropriate technology
development and dissemination, most of the proposals, particularly for international action, have so far remained on
paper. One of the reasons for this is perhaps the controversy concerning the establishment of international institutions and research institutes of the type established under
the auspices of the Consultative Group on International
Agricultural Research. Is it better to carry out research
and development and information dissemination in existing
national institutions or in new international institutions?
This remains a moot question. There are arguments for and

---

[1]Chief, Technology and Employment Branch, International
Labour Office, Geneva.

against both types of institutions. In favour of the
international institutes, it can be said that in many cases
the same basic design of hardware is applicable with no
more than simple refinements to a number of developing
countries. At times it is also argued that an international
institution can attract talent without fear of losing it to
local universities and institutes. Thirdly, the record of
many national institutions in developing countries over the
past several years has been rather poor in terms of results.
In theory, a national "centre of excellence" could be transformed into a suitable international institution rather than
creating one de novo. Yet, there may be difficulties in
this transformation process.

There are those, like the authors in this volume, who believe that research should be carried out within national
institutions in developing countries in order to ensure
that effective links are maintained between laboratories
and users. A strategy of reorienting R and D towards the
rural and small-scale urban sectors adopted in this volume
makes it even more necessary that research is undertaken on
the spot so as to adapt the hardware and software to local
conditions and to capture learning effects of research locally.

If the major objective of action on appropriate technology
is to ultimately build national technological capacity in
the developing countries, then clearly the role of international action can be no more than sensitisation, financing
of R and D, and other catalytic action in support of existing or potential national efforts. The following key
questions need answering before a pronouncement can be made
on the form and content of an institutional framework for
international action:

(a) what are the criteria and an operational definition of
    technologies appropriate for a basic needs or any other
    type of development strategy?

(b) what role do existing national and sub-regional institu-
    tions, particularly in developing countries, play in the
    development of appropriate technologies? How can exist-
    ing institutional capacity in developing countries be
    strengthened?

(c) what mechanisms, if any, exist within the UN system to

ensure coordination of activities of different agencies
within the framework of a unified policy for the UN
system as a whole?

(d) what mechanisms exist for networking, information dissemination and promotion of research and development,
and how effective are they?

(e) in the light of the above, is there a case for a new
international mechanism for the promotion of appropriate
technology?

The six chapters contained in this book provide at least
some answers to these questions.[2]

Part I of the book deals with concepts, definitions and
strategies. In Chapter 1, Nicolas Jéquier spells out some
relatively unfamiliar criteria which should be taken into
account in guiding the selection of appropriate technologies.
Jéquier adopts a rather narrow definition of appropriate
technology to mean small-scale technology that often has its
origins in the traditional methods of production. He presents a number of criteria of appropriateness, e.g. cost,
risk involved, modernity, individual or collective nature,
and single or multipurpose character of technology.
According to Jéquier, problems of non-economic nature, that
is, sociological and institutional constraints, to the
widespread adoption of appropriate technologies are more
important than the purely economic considerations.

My paper (Chapter 2) assumes that the developing countries
are (or should be) concerned with a rapid fulfilment of the
basic needs of the poorer target populations through increasing national and collective self-reliance within the
framework of a new International Economic Order. The links
between the concepts of appropriate technology on the one
hand, and basic needs approach to development on the other,
are established by outlining a technological content of a
basic needs approach, by examining the demands that this

---

[2]Earlier versions of two chapters, namely, by Amulya Reddy
and by the author, appeared as ILO World Employment Programme Working Papers.

approach might make on decentralisation of production
structures, and by considering the political and administrative requirements of technological self-reliance of
developing countries.

Whatever the types of development strategy, no objectives,
targets or instruments of action are likely to be implemented without the existence of adequate institutional
mechanisms. Part II of the book therefore examines the
existing institutional capacity at the national and international levels.

Amulya Reddy, in Chapter 3, argues that national capabilities in developing countries do exist in the form of science
and technology institutions. However, at present these institutions are not harnessed for the development and dissemination of appropriate technology on any appreciable scale.
Very few of these establishments have any direct contact
with the users through field stations or extension centres.
This situation has contributed to the neglect of the needs
of the small-scale enterprises whose own resources are too
limited to finance research for upgrading of technologies.

Two chapters are devoted to the international dimension of
the problem.

The activities of the UN system on appropriate technology
are examined in Chapter 4. Willem Floor reviews the on-
going activities of the UN agencies and comes to the conclusions that (a) appropriate technology with special reference to anti-poverty and basic needs approaches, forms
only a small part of the UN activities on science and technology, (b) considerable overlapping of activities of
different organisations occurs owing to lack of any systematic joint planning of programmes and (c) coordination is
needed mostly at the country project level.

Frances Stewart, in Chapter 5, reviews existing international mechanisms (both within and outside the UN system)
and the proposals for new mechanisms dealing directly or
indirectly with appropriate technology. She concludes that
the existing mechanisms do very little on appropriate technology. Significant international action to improve information collection and dissemination and to promote
appropriate R and D on appropriate technology, is therefore
needed to supplement national efforts.

In the light of the preceding review of existing national
and international mechanisms, Chapter 6 presents a blue-
print for global action for the promotion of appropriate
technology, through the establishment of a new mechanism.
Extensive consultations in a cross-section of countries
and at the headquarters of various UN and other intergovernmental organisations were held by Paul Marc Henry,
Amulya Reddy and Frances Stewart regarding the feasibility
of a new mechanism. The action programme proposed in this
chapter reflects the majority opinion of the individuals
and institutions in the developed as well as developing
countries. It is also quite timely in view of the forthcoming UN Conference on Science and Technology for Development (UNCSTD) to be held in August 1979.

A programme of action for UNCSTD is currently being prepared by the UNCSTD Secretariat on the basis of proposals
from countries, regions, UN organisations and the scientific community. According to the Secretary-General of UNCSTD,
this programme of action is likely to reflect three "end
products" of UNCSTD, namely: (a) appropriate mechanisms for
a harmonised UN science and technology policy; (b) a horizontal mechanism for the exchange of experience among developing countries concerning technology development and
transfer; and (c) a few specific pilot projects on topics
of high priority. The blue-print for action proposed in
Chapter 6 is particularly relevant to the horizontal cooperation among developing countries.

The new international mechanism is intended to overcome the
imbalance of work on technology between developed and developing countries, and to promote sound national technological capability for generating indigenous technologies best
suited to the prevailing conditions of developing countries.
This mechanism is proposed as a flexible entity like the
currently operating consultative groups. It is not, however, intended as a new institution. Although it will resemble some consultative groups, it has a number of its own
distinctive features. Firstly, unlike the consultative
groups, it will not be dominated by the donors from the developed countries or from the developing countries for that
matter. Greater participation of developing countries is
anticipated. Secondly, it is proposed to be formally outside the United Nations System. Nevertheless, it will need
to be closely associated with the different UN organisations
currently engaged in work on science and technology.

Thirdly, it would not be responsible for coordinating UN activities in the field of appropriate technology. This is a task to be undertaken by appropriate bodies within the UN system.

PART I
Concepts, Criteria, Strategies

Chapter 1
APPROPRIATE TECHNOLOGY: SOME CRITERIA

N. Jéquier

INTRODUCTION

Theories about economic development are subject to the same whims of fashion as women's clothes: what appeared beautiful one season seems a little out of tune the next and distinctly inappropriate three years later. There are good reasons to believe that our concepts, ideas and policies in the field of appropriate technology are in need of a little re-tailoring, if not of major re-designing. To put things in a more technical perspective, we must now make the jump from the "first generation" to the "second generation" in appropriate technology.

The first generation can be characterised by the paramount importance of moral and ideological considerations in the debate about development styles, by the seminal role of a small number of marginal groups in bringing these issues to the forefront of development thinking, and at the technological level, by the experimental nature of innovations in hardware. This first phase we have known in the last few years has been extremely important, but it has probably reached its natural limits.

The second generation which is now opening up raises a number of new and largely unexplored issues. One is that of institutionalisation or "de-marginalisation" of appropriate

----

[1]Principal Administrator, OECD Development Centre, Paris.

technology. Another is that of effective linkages between
the innovation system in appropriate technology and the
financial and investment system. A third is that of developing national and international technology policies
focusing specifically on appropriate technology. And the
fourth is the role of appropriate technology in meeting the
basic needs of the hundreds of millions of poor people in
the less developed countries.

The purpose of the present paper is not to outline in any
detail the scope and contents of this second generation
strategy in appropriate technology, but rather to explore
one of the critical issues in this debate, namely that of
the appropriateness of technology. One assumption here is
that the shape of this strategy will be determined to a
certain extent by the criteria and yardsticks according to
which a technology is defined as appropriate or inappropriate. Another assumption is that until now most of the
yardsticks and criteria of a technology's appropriateness
have been economic, and one aspect that has been particularly emphasized is that of a technology's effects upon
employment. These criteria are important and the methods
of economic analysis have become extremely sophisticated.
However, without wanting to belittle the role of economic
yardsticks and notably of their employment-generating component, it is well to note that economics is not everything,
and that success in the innovation process depends not only
on the economic attractiveness or appropriateness of a
particular technology, but also on a number of social,
cultural and technical factors which are perhaps more difficult to pinpoint and quantify, but which are at least as
important.

The present paper will try to identify some of the less
widely used criteria. Our hypothesis is that by considering
such other criteria, it may be possible to design technologies which are in effect more appropriate than those whose
main justifications are economic. What follows here is not
a systematic methodology, but rather a presentation of potentially relevant criteria illustrated wherever possible by
specific examples. For obvious reasons, it would be somewhat unrealistic to attempt to rank them according to importance: some may be particularly relevant in certain
specific areas, and irrelevant in others, and the discussion
which follows is intended merely to present some elements
that might be taken into account both in the design of new

appropriate technologies, and in the definition and implementation of technology policies focusing specifically
on appropriate technology.

At this stage, it may be useful to clarify a certain number
of points. The first is that the search which is currently
going on both in the industrialised and the developing
countries for more appropriate forms of technology is
probably not an ephemeral fashion, as many critics of the
appropriate technology movement would claim, but rather the
manifestation of deep-rooted social and political changes
which are only just beginning to be translated into new
types of technology and novel approaches to innovation.
Appropriate technology is here to stay, but it should also
be realised that the scope and number of successful innovations in appropriate technology is for the moment still too
limited to serve as a convincing and viable alternative to
the types of technology we have today. The situation is
somewhat similar to that of the automobile in 1890: this
new technology looked very promising, particularly to those
who had developed it, but it was not yet a competitive substitute to the railway and the horse-drawn carriage.

The second point is that appropriate technology should not
be viewed as a second-rate technology. Designing a good
appropriate technology, be it an inexpensive and reliable
water pump, a long-lasting roof for slum dwellings or a
truly efficient oxcart, is in many ways just as complex
and challenging from the conceptual point of view as any
modern industrial innovation. What is more, the diffusion
of such appropriate technologies is in some respects much
more complex than the diffusion of modern consumer goods or
new industrial production processes: social and cultural
resistances are much stronger, the income level of those
who stand to benefit from such innovations is usually very
low, and the market forces which could stimulate innovation
tend to be rather inarticulate. Appropriate technology is
not, and should not be viewed as a second-best solution.
Conversely, neither should its role be over-estimated:
appropriate technology is not a universal substitute for
the conventional modern technology. Appropriate and modern
technologies are complementary rather than contradictory,
and the emphasis given to the former does not and should
not rule out the use of the latter in those cases where
they are particularly well adapted to local situations.

A third point which deserves some clarification here is the
relationship between appropriate technology and the satisfaction of basic human needs. In the industrialised countries, a sizeable proportion of appropriate technology groups
are working on the development of innovations which focus on
the better utilisation of scarce natural resources, the
transition to renewable sources of energy and the minimization of technology's negative impact on the environment.
By contrast, most of the groups working in the developing
countries, as well as those groups in the industrialised
countries which have given particular emphasis to the problems of developing countries, tend to view appropriate
technology as the main tool in meeting the basic needs of
hundreds of millions of poor people who have largely been
left out of the development process. The picture presented
here is no doubt somewhat caricatural, but it does suggest
the existence today of two big "families" in appropriate
technology: the "industrialised country family", which is
concerned largely with environmental questions, natural
resources and the technologies of the post-industrial
society, and the "developing country family" which gives
much more attention to the problems of poverty, social
equity, employment and basic human needs. There is much
overlapping between these two families, and their genetic
and cultural backgrounds are not fundamentally different.
Furthermore appropriate technologies developed by one of
these two families can in many cases be considered as
equally appropriate to the concerns of the other: this is
the case for instance of many technologies in the field of
energy, as well as in agriculture. When looking, as this
paper does, at the criteria of a technology's appropriateness, it is important to bear in mind the differences between these two "families" and to realise that what is
appropriate to one of these families will not necessarily be
considered in the same light by the other. What we will do
here is try to identify a certain number of criteria of
appropriateness which are common to both families.

SOME CRITERIA OF APPROPRIATENESS

"Systems-Independence" of New Technology

The first criterion we shall examine here might be called
the "systems-independence" of a new technology. This some-

what obscure term can perhaps best be explained with the
help of a few examples. The first is that of the high
yielding varieties (HYV) of rice developed by the
International Rice Research Institute and which are one of
the mainstays of the Green Revolution in the humid tropics.
One central feature of this new technology is its dependence
on a wide range of supporting services and technologies.
First on water: these HYVs require irrigation (rain-fed
HYVs are still at the development stage), which means that
they can only be used by farmers who are already growing
irrigated rice (and these are usually the more affluent
farmers, or those living in the most developed regions of
the less developed countries). The second is on fertilizers
and pesticides. The third is on mechanisation, which is
needed in order to meet the high demand on labour. And the
fourth is on supporting services such as drying equipment
(crops maturing during the rainy season cannot be sun-dried,
but require diesel-powered driers). There is no doubt that
these HYVs are in themselves a remarkable piece of biological technology, but their diffusion is contingent upon
the availability of a wide range of supporting services and
subsystems. In other words, it is a highly "systems-
dependent" type of technology. This in itself is not
necessarily a drawback, but it clearly limits the number
of potential beneficiaries, notably among the poorest
farmers in the least developed regions.

A second example is that of the methane-gas plants developed
among others in India. Here again, this is an extremely ingenious technology, but like the HYVs of rice, it is very
system-dependent: the farmer who wants to operate such a
domestic plant must have at least two or three cows to provide the raw material, he must have an easy supply of water,
and he needs a certain amount of land around his house, both
to install the machine and to dispose of the fermented dung
as a fertilizer. The systems-dependence of this particular
technology limits its attractiveness to a rather narrow
group of farmers, namely those who are rich enough to own
several cows and have their own house and some land, and at
the higher levels of income, to those who are not tied into
an integrated sewage disposal system and into the electricity
distribution network.

Another example is the case of powdered milk for babies.
Despite claims to the contrary, this 19th century technology
is extremely appropriate (one may think here of the ease of

storage and distribution) and it played a major role in
bringing down infant mortality in the countries which are
today highly industrialised. But this technology is highly
systems-dependent: clean water must be available for mixing
with the milk powder, and cheap fuel is needed to sterilise
the bottles and the water, two conditions which are seldom
fulfilled in very poor communities. If these two supporting services are absent, the inherently appropriate technology of powdered baby milk may become very inappropriate.

These few examples suggest that some types of technology
are by nature more systems-independent than others. This
is not to say that they are inherently better or more
appropriate than technologies whose successful diffusion
depends upon the presence of an important supporting infrastructure (education and training facilities, extension
services, provisions for credit or new forms of social
organisation). Nor does it imply that systems-independence
is a feature that can somehow be engineered into a new
piece of hardware at the design stage. It does however
point to the fact that in the innovation process, the key
to success lies not only in the intrinsic qualities of a
particular technology, but in many cases in the presence of
a wide range of supporting services. This seems to be
particularly true in the case of technologies aimed at
meeting basic needs. When trying to evaluate the appropriateness of a particular technology, one should consider
not only the technology as such, in the form of a clearly
identifiable piece of hardware, but also all its supporting
software. It is also important to realise that in many
cases, a trade-off may occur between hardware and software:
if the supporting package is highly effective but the hardware of average quality, innovation can be easier to achieve
than in the case of an excellent piece of hardware without
any such supporting services.

Image of Modernity

People buy a product or use a specific technology because
it is economically attractive, socially useful or technically appropriate; but they are also influenced by its
symbolic value, and by their perception of the product's
modernity. In developing countries imported products and
technologies very often have a better image than local
products, and national trade policies which seek to limit

imports in order to promote indigenous industries may
ultimately contribute to enhancing the image of these
foreign products - because of their growing scarcity.
Development experts and appropriate technology proponents
are quick to deplore and sometimes to deride this attrac-
tion for things foreign, which is seen as a manifestation
of cultural imperialism on the part of the industrialised
countries and of social alienation on the part of the
developing countries. What is not always realised is that
this positive image of foreign and modern products or
technologies, which entails a number of drawbacks, can
also be an important feature in the process of development.
Rather than try to pass a judgement - and the judgement is
invariably negative - it may be better to try to exploit
this cultural feature of the innovation process. Two
examples can be given here to illustrate both the positive
and the negative approach.

The first is that of the small-scale sugar plants now
operating in India. These small plants designed by the
Planning and Action Research Institute in Lucknow embody
what by any counts can be considered as a particularly
appropriate technology (high labour-intensity, competitive
production costs, low investment per workplace, high
quality of end product). Of interest here is the fact that
the designers perceived very clearly that consumer pre-
ference was shifting away from traditional sweetening
agents like gur and khandsari to the more modern (and
nutritionally perhaps less appropriate) white crystal sugar.
As a result, these small-scale sugar plants were designed
specifically to produce a high quality white crystal sugar
similar to the one made in the large-scale modern plants.
This was undoubtedly one of the important if unrecognized
factors in the success of this major innovation in
appropriate technology.

Contrast this with the inexpensive high protein instant
food developed by the subsidiary of a large multinational
American food firm for sale in Latin America. The success
of this new cereal was evident in a number of countries
(a testimony, incidentally, to the ways in which large
multinationals can be encouraged to develop appropriate
technologies), but in one particular country the innovation
was a complete failure. Subsequent investigations showed
that this failure did not result either from the quality
of the product or its lack of attractiveness in terms of

taste, but rather to the fact that it was deliberately
marketed as a poor people's food.

It might be suggested here that in many cases one criterion
in the design of appropriate technologies for basic needs
is that of modernity. This is not to say that the aim of
the poor people in the developing countries should be to
imitate the life style and consumption patterns of the rich
in the industrialised countries or of the local elites in
their own country. Even less are we trying to suggest that
the only path to development is that followed by the
countries which are today highly developed. The fact is
however that in many instances the drive which motivates
people to improve their living conditions can be satisfied
by technologies which, in addition to their intrinsic
qualities, also carry with them the image of modernity.
One of the very difficult problems here is not to transpose
from abroad new technologies which are both modern and
appropriate, but to design locally technologies which are
consonant with the society's culture, values and resources[2].

Individual Technology Versus Collective Technology

Many of the technologies required to satisfy basic human
needs are of a collective or community nature: this is the
case with water supply systems, sewage systems, electricity
distribution services, and many others. Others can be
either collective or individual: in the case of housing for
instance, there is a choice between individual dwellings or
apartment complexes. And some technologies are essentially
individual in nature: this is the case of most durable
consumer goods. Clearly this distinction between individual
and collective technologies is somewhat arbitrary, and in
most cases there is an element of each mode in every technology: an automobile embodies a more individual technology
than a railroad or a bus, but its operation requires certain
collective services (e.g. roads, maintenance services, etc.).
An electric stove is a more individual technology than a gas
stove tied to a village-level gobar plant, but it is clearly
more collective than a butane cooker using rechargeable con-
tainers.

---

[2]See Chapter 3.

The distinction made here between individual and collective
mode does not imply that one mode is inherently superior to
another, but rather suggests that this may also be an important criterion in determining both the appropriateness
of a technology and the directions in which the innovation
process might be channelled. It is also one of the criteria
which tends to be overlooked in the traditional economic
evaluations of a technology's appropriateness.

This can be illustrated in the case of the diffusion of
low-cost water purification and supply systems in Thailand
and the Philippines. The technology, developed in Thailand,
is both simple and appropriate: turbid water from canals or
rivers is purified in a village plant with a filtering
medium made from agricultural wastes (rice husks or coconut
husks), and small quantities of chlorine are later added to
the purified water. This technology requires a basic
community infrastructure: arrangements must be made for
sharing the costs of operation, and one or two persons must
be entrusted with the running of the plant. This technology
was tested in a number of villages both in Thailand and the
Philippines, and interestingly enough was much more successful in the Philippines. In Thailand, where social allegiance
is more to the extended family than to the community, a significant proportion of users dropped out of the scheme
(which meant that the cost of water for the other families
had to be increased), the villages were reluctant to pay the
salary of the plant's operator, and there were often major
difficulties in allocating the limited quantity of purified
water to the participating families. In the Philippines by
contrast, the operation of such plants proved to be much
easier, largely because of a long-standing tradition,
nurtured perhaps by religion, of community participation.

The same problem can be found in the field of agricultural
mechanisation. In many cultures, co-operative forms of
organisation of the mechanisation process are extremely
difficult to implement, and the more appropriate strategy
may be to introduce small machines that can be individually
owned and operated, rather than the larger machines that
require a co-operative form of organisation. And if large
machines (e.g. bulldozers, large tractors, etc.), are ab-
solutely required, for instance because of the particular
nature of the terrain or the structure of the land holdings,
it is possible to design individual modes of operation for
machines which in other cultures would more appropriately

be used on a collective or co-operative basis. For instance
by financing tractor hire systems (the machine is owned by
an individual, but farmed out by the hour or day, with a
driver, to the farmers who require its services).

Slum rehabilitation is another typical field where choices
can be made between collective and individual modes of
innovation. Relocating slum dwellers in high-rise apart-
ments can be successful (as in Singapore, a Chinese
Confucian culture), but in many cases, the new dwellings
gradually degrade into high-rise slums. But very often,
with small individual houses which leave room for improvement and enlargement, the process of relocation or
rehabilitation often leads to further improvement and development. One very interesting example in this respect
is what happened in the city of Cali in Colombia, where
the provision of basic services, cheap credit and technical
extension services were instrumental in the upgrading of
slums into middle class areas.

What these examples suggest is that the mode of operation
of a new technology - collective or individual - is often
a crucial element in the success or failure of an innovation.
When designing a technology, account must be taken of the
culture of the group for which this technology is intended,
and this criterion is probably a much more important
determinant of success or failure, and hence of a technology's appropriateness, than generally suspected.

Cost of Technology

Few people would disagree with the statement that one of
the essential features of an appropriate technology is its
low cost. This is particularly true when it comes to
technologies aimed specifically at meeting basic human
needs. However, while there may be a general agreement
about this worthy principle, it is well to note that the
issue is in fact much more complex than generally suspected,
and conflicting approaches to the cost elements account for
a large share of unsuccessful innovations in appropriate
technology.

Just to illustrate the problem, let us return for a moment
to the case of the Indian mini sugar mills mentioned earlier. In terms of production costs, it can be observed

that the kilo of sugar from these mills is slightly cheaper
than that from a modern large mill. This cost advantage
however is not due only to the inherent efficiency of this
process, but also to a number of exogenous factors. One is
the lower taxation rate. Another is the lower transportation cost of cane from the plantation to the mill and of
refined sugar from the mill to the village store. And the
third element is the lower depreciation allowance for
capital investment. These are factors over which the mill
owners have little if any control, and what is more, they
can change rather abruptly, thereby modifying the underlying economic assumptions about this technology's
efficiency.

If one considers things in a long-term perspective, the
cost equation becomes even more complex. The small plants
employ a lot of labour, and it can be expected that the
general rise in the country's income level coupled with a
deliberate policy of increasing the income of the lowest
paid workers, and the growing unionisation of the labour
force, will contribute to raising the labour costs of the
small mills somewhat faster than those of the large modern
mills. In other words, their costs will increase rather
rapidly, and the economic justifications for such a technology may not be as compelling ten or fifteen years hence
as they are today. This is particularly true of appropriate technologies such as this one which stand so to
speak at the periphery of the modern industrial sector and
which have to compete on the open market against the
technologies, organisational power and financial resources
of this sector. The problem is undoubtedly somewhat
different in the case of appropriate technologies aimed
specifically at meeting the basic needs of a very poor
population: costs and benefits are rather difficult to
measure in purely economic terms, ideological and moral
factors are particularly important, and social efficiency
is usually more meaningful than private efficiency.

Here it may be useful to draw a page from the accounting
books of modern industry, and notably from service-oriented
industries such as telecommunications. Accountants of the
national telephone authorities are now accustomed to measure
the cost of any piece of equipment not simply in terms of
purchase price but in terms of total life cost of the
product. As a result depreciation charges, maintenance
costs, repairs and expected useful life become more im-

portant in the calculus of efficiency than the initial and
somewhat narrow production or purchase cost. In the same
way, it may be useful here to envisage some more sophisticated accounting methods for measuring the value to envisage
some more sophisticated accounting methods for measuring the
value of appropriate technologies. This concept of total
life cost would for instance clearly bring to light the fact
that the relatively low reliability of many appropriate
technologies makes them less attractive than a short-term
estimate of present operating costs would suggest (and this
indirectly explains the market failure of many apparently
appropriate technologies) and would pave the way to a much
more serious attention to this problem of reliability and
durability.

Technologies aimed at solving the major problem of basic
human needs are presumably aimed at the poorest people.
Leaving aside for the moment the issue of social acceptability, it is clear that one of the critical elements is that
of cost, or if placed in another perspective, that of the
price charged to the user or consumer. Evidence seems to
suggest that a large proportion of the low-cost appropriate
technologies now available to meet certain basic needs (in
housing, food, shelter, health, etc.) are still much too
expensive for the very poor. In the same way that one of
the basic problems in nutrition is not the lack of food,
but the inability of people to pay for the food they need.
One can either try to raise the income of the poorest
people in order to allow them to purchase what they need
to meet their basic needs, or else develop technologies or
products which are much lower in cost and price than is the
case today. The cost decreases which are required here are
not of a few percentage points, but of one order of magnitude at least, and this presents our innovation systems
with a major challenge which in most cases is still far
from having been met.

One of the difficulties here is that when it comes to
certain types of basic needs the innovation system in
appropriate technology does not seem to be subject to the
same experience curves that can be observed for instance
in modern industry. Nor does it seem always able to benefit
from the same advantages of scale. The experience curve
theory states that production costs per unit decrease in
proportion with the cumulative number of units produced.
The rates of decrease vary of course considerably from one

industry to the other, but the general pattern can be
verified empirically from a wide range of sectors (e.g.
automobiles, computers, aircraft, machine tools, etc.).

In the case of appropriate technologies aimed at meeting
basic needs these two phenomena of experience curve and
economies of scale may operate in a rather less effective
way. One possible reason for this is that appropriate
technology, particularly when it is conceived of as a tool
for self-help, often represents a "one-of-its-kind" type
of technology. The farmer who is taught to build a more
appropriate water storage tank or to use a new inexpensive
material for roofing his house, will in most cases do the
work that is required for his own needs, and stay at that.
In the same way, one can help thousands of poor people to
build a latrine and sewage pit, but if there are economies
of scale, they can be found in the teaching and extension
process, rather than in the production process itself.

This method of operation, characteristic of self-help projects, is psychologically very important in that it shows
the beneficiaries that they can improve their lot and
master at least some aspects of their life. But it also
means that every self-made piece of equipment is burdened
with the additional costs (and often the technical deficiencies) which are associated with the first product of
a batch.

The existence or absence both of learning effects and
economies of scale is probably an important criterion in
selecting technologies and in initiating development projects. To take a somewhat hypothetic example, it might be
worth putting more emphasis on the development of family
size low-cost water purification units which can easily be
transported from a regional manufacturing unit, than on the
larger, untransportable (and therefore difficult to duplicate) village-level plants built by the villagers themselves. The question here is not only whether a particular
technology is inherently subject or not to the effects of
learning and economies of scale, but also whether the mode
of production, the system of diffusion of innovation and
the method of operation of the extension service is not in
itself a more important variable in this respect than the
technology as such. And the criterion of appropriateness
which might be suggested here is that of sensitivity of a
technology (including all the software which accompanies it)

to the effects both of learning and economies of scale.

Risk Factor

Any innovation involves a certain amount of risk: for the
industrialist who launches a new product, for the farmer
who purchases a new machine or tries out a new crop, for
the consumer who buys a new house or moves to a different
job. The interest of any innovator, whatever the level of
sophistication and complexity of the innovation, is to try
to minimize these risks. In the case of a new product, the
consumer will turn to a well-known manufacturer with a high
reputation, and the farmer will closely watch others who
are experimenting with a new crop or a new way of doing
things.

In the case of appropriate technology, three general observations can be made. The first is that this problem of risk
is largely if not totally absent from discussions about the
choice of technology, the innovation mechanisms required to
promote innovation, or national development policies. The
second is that in the case of technologies geared specifically to the poorest income groups, the risk of any particular innovation is inherently much greater than in the case
of the population at large: the lower the income, the
greater the relative risk for a given innovation or technology. And the third is that a new technology (and most
appropriate technologies belong to that category) almost
always involves greater risks than a well-established
technology, even if it is economically and technically
much more attractive. Risk is a very important criterion
in determining the appropriateness of a technology, and it
is little more than a statement of the obvious to say that
the most appropriate technology is also the one that involves the minimum amount of risk for the user.

If this is accepted as a valid criterion of appropriateness,
it is possible to start designing technologies and their
supporting services and subsystems with this objective of
risk minimization in mind. Clearly, all designers of new
appropriate technologies somehow have this objective in
their mind, even if only in an implicit way, just as an
engineer does in any modern industry. But what is not
always fully realised is that the level of risk - economic,
social, cultural or technological - involved in any particular new technology depends not only on the inherent design

qualities of this technology (this might be called the
internal risk) but also on the ways in which it fits into
the local production system, the local culture and the
available supporting services (this might be called the
external risk).

Let us consider for a moment the new types of low-cost
transportation vehicles developed by several research
centres for the rural areas of the developing countries.
These small, rugged and inexpensive vehicles (such as those
of the International Institute of Tropical Agriculture in
Nigeria or of several industrial enterprises in the Philippines) are in many ways very appropriate. But what must be
realised is that both the internal and the external risks
involved are considerably greater than in the case of conventional vehicles. The internal risk stems from the
novelty and somewhat experimental nature of this technology.
The same problem can of course be found in all sectors of
modern industry: purchasers of the first models of a new
automobile for instance are almost always plagued with the
teething troubles characteristic of any new vehicle. As
for the external risk, it results in large part from the
fact that the supporting services that might alleviate the
internal risk (i.e. maintenance and repair services,
supplies of spare parts, etc.), are simply not available,
at least for the moment.

Internal and external risk tend to operate in a synergistic
way, which can be either positive or negative. If the
design of a new product or a new technology is sufficiently
good to reduce the internal risk to a minimum, it tends by
way of consequence to reduce at the same time the external
risk (a very reliable machine for instance needs only
minimal supporting services). Conversely, a high degree of
internal risk puts very heavy burdens on the supporting
services, a phenomenon which in turn contributes to re-
ducing the effectiveness of these services, and hence to
increasing the external risk.

Another important factor to consider here is that the
internal and external risks of any new technology tend to
become proportionately much greater when attempts are made
to introduce at the same time several other inter-related
technologies. To put things in a rather extreme form, one
might say that the risk involved in an innovation increases
not linearly with the number of new technologies involved

in that innovation, but as the square of that number. This
can be illustrated with the case of an integrated energy
supply system for villages now being tested in at least two
very poor countries. The idea here was to satisfy all the
energy requirements through a set of apparently very appropriate technologies: a solar pump for drawing water from
the aquifer, bio-gas plants for the supply of cooking fuel
and the disposal of animal dung, and windmills for the production of electricity. All three technologies are rather
experimental, and largely untested in that particular social
and economic environment. Quite apart from the fact that
the cost of these pilot projects is totally out of proportion
with the resources of the local communities, it can be observed that none of these technologies is operating satis-
factorily. For instance, the new grain mill introduced to
make use of the available electricity produces a flour of a
rather different texture than the one made by manual
crushing, and no one wants to eat it; as for the solar
pump, its main effect has been to multiply water consumption
by a factor of ten, at least during the time when it happens
not to be laying idle because of technical problems. The
whole innovation having been conceived as a system of interdependent technologies, the failure of any of them directly
affects all the others. In such a case the overall risk of
failure of the whole system is not equal to the sum or the
average of the risks characteristic of each component, but
to the rate of risk of each component multiplied by the
rate of risk of all the others.

Without going into the rather complex issue of risk measurement, it can be observed that this notion of risk is not
only economic (e.g. what are the chances that a new piece of
equipment will perform as effectively as anticipated?), but
also technical (e.g. what is the risk of technical failure
over a given period of time?). Furthermore, risk also depends on social and cultural factors. Certain new technologies, because they are well tested, may have a very low
technical and economic risk, but may ultimately fail completely because the social and cultural factors were neglected. This is particularly frequent in the case of
technologies related to food production and consumption,
and also but to a lesser extent, to technologies related to
housing.

What these few examples suggest is that risk is one of the
important determinants in the appropriateness or inappro-

priateness of technology. And as we have tried to show,
the risk can be either internal or external, it is not only
economic and technical, but also social and cultural, and
account must be taken of the fact that the risks of an
innovation which involves a large number of components or
subsystems tends to be considerably greater than in the
case of an innovation embodying a single new technology.

If the ideal appropriate technology is often the one which
involves the minimum risk to the users (particularly in the
case of people who belong to the poorest and most under-
privileged social groups), it is also worth pointing out
that high risk in itself is not always necessarily a
negative feature. In the industrial sphere in particular,
the high risk of an innovation can be partly balanced if
not totally outweighed by potentially very high economic
rewards, and it can be observed that dynamic young entrepreneurs are often attracted by ventures where both risk
and potential rewards are far above average.

It should also be noted that if low risk is given an
exaggerated importance in determining the appropriateness
of a new technology, there is a danger of stifling the
innovation process and, when different solutions are
available to meet a particular need, of favouring systematically the well-tested technology, which in many cases
is an imported technology rather than an indigenous innovation. Since one of the main aims of the appropriate
technology movement is precisely to foster greater local
self-reliance and alleviate some of the most conspicuous
problems of what might be called the addiction to foreign
technology, it would be inadvisable to look at the risk
factor as an important criterion of appropriateness.[3]

Evolutionary Capacity of Technology

One of the many reasons why traditional technologies are
not as competitive as modern technologies is that they are
relatively static. This is not to say that they do not or
cannot benefit from any innovations. But the process of
innovation, when it does take place, is generally slow, and

---

[3]For a discussion of self-reliance, see Chapter 2 below.

in most cases too slow to keep up with the pace of innovation in the modern sector. In the case of appropriate technology, which is often an intermediate between traditional and modern technology, the patterns of innovation are often closer to those that can be found in the traditional sector than to those of the modern sector. In other words, many of the new technologies developed by appropriate technology groups do not have a sufficient evolutionary capacity, and innovation tends to be conceived of as something that happens once and for all.[4]

The problem here lies not so much in the nature of technology as in the culture and philosophy of the appropriate technology movement. What seems to be happening in effect is that its reference framework is the innovation system in traditional technology, rather than the innovation system in modern technology. Because of this orientation, there is a tendency to overlook the inherent dynamism of innovation in the modern sector, and to design technologies which are in many cases more appropriate than the traditional technologies they seek to replace, but which in the long run stand little if any chance of becoming competitive relative to modern technology.

Another reason for this is that appropriate technology, at least in the current sense of the word, is a fairly new concept. Developing new sources of energy for the villages, inexpensive water supply systems, more efficient small-scale plants for rural industries or new types of simple agricultural machinery is in itself a very complex task (contrary to what is usually believed) and partly for this reason, it is rather difficult for the innovators in appropriate technology to perceive that what they have developed, often with great difficulty and using much ingenuity, is transient and provisional.

Appropriate technology, if it is to have an evolutionary capacity, will in many cases (notably when it deals with production technology as opposed to self-help technology) evolve into a larger scale and more capital-intensive type of technology, which is precisely the type of technology

---

[4]See Chapter 3.

(with its associated philosophy, work patterns and social
relations) to which the appropriate technology proponents
are seeking a valid alternative. Illustrative of this
problem is the story of the small-scale egg tray making
machine developed by the Intermediate Technology Development Group. Attempts to improve this ingenious design
resulted in the development of a better machine which
employs half as many people as the one it replaces. This
may well be a small example, but it is probably rather
more typical than one would suspect. This indirectly
suggests that the evolutionary capacity of a technology is
not something inherent in the technology itself. Rather
it is a feature built into the philosophy of those who
innovate. And in the case of appropriate technology, there
may well be some sort of a mental block against a system of
permanent innovation which might lead to larger scale and
less labour-intensive types of technologies.

Symbolic of this problem is the major focus of appropriate
technology strategies on the issue of employment, and
specifically on one particular aspect of the employment
equation, namely the cost per workplace. This concept of
cost per workplace has been extremely fruitful, and points
to some very fundamental issues, notably to the fact that
societies with very limited resources cannot afford to
invest large sums for each job created if they are to
improve the overall employment situation. Without wanting
to belittle the major importance of this criterion of
appropriateness, it is interesting to observe that considerably less attention has been given to the other side
of the coin, namely output per employee. If output is low,
incomes accruing to the workers will also be low, and the
only way to increase incomes is to ensure that the productivity of these appropriate industrial or agricultural
processes will also increase.

Rather than use investment cost per workplace as one of
the most important criteria of a technology's appropriateness, it might also be worth considering output per worker,
and more specifically output over a long period of time.
Indirectly this would help to measure the evolutionary
capacity of a new appropriate technology, and serve as
feedback to the innovation process.

Consider for instance the case of a country with a per
capita income level in the range of $200. A new industrial

plant with an average investment per workplace of $500 is
in many respects more appropriate than a much more capital-
intensive plant with an investment per workplace of
$20,000. These two plants will in all probability have
very different levels of productivity per worker. If the
output per worker of the first plant is around $1000, this
may be fine for the time being, but the crucial question
here is whether this smaller scale and more appropriate
technology can, in some way, be upgraded so as to push
output per worker to $2000 in five or ten years time (the
figure being measured here, of course, in constant terms).
If output per worker ten years later is still likely to be
no more than $1000, then it is probable that this is not as
appropriate a technology as initially thought. And the
main result of this innovation will have been to create
another type of ghetto, not entirely unlike that which
characterises the traditional sector in many developing
countries.

There are good reasons to believe that one of the criteria
of a technology's appropriateness is its evolutionary
capacity. However, because of the complexity of measuring
what constitutes a technology's evolutionary capacity (the
term may be clear, but its practical measurement is extremely complex, and probably very subjective as well), it
might be easier and more simple to use a substitute yardstick, namely output per worker over a relatively long
period of time. If this output is likely to remain static,
the technology is probably less appropriate than initially
envisaged. One of the functions of this yardstick however
is not simply to make rather theoretical long-term com-
parisons, but rather to bring innovators in appropriate
technology to think not only in terms of today's needs and
resources, but also in terms of building up a system of
permanent innovation in appropriate technology.

Single-Purpose and Multi-Purpose Technology

Another criterion of appropriateness which would deserve
closer attention is that of the range of applications of a
technology. Is a "multi-purpose" technology more appropriate than a "single-purpose" technology? This concept
can be illustrated by the case of small-scale agricultural
machinery. One of the most successful and appropriate types
of machine now being diffused on a large-scale in several

Southeast Asian countries is a small power tiller which
can be used for a number of different purposes: designed
initially for tilling, it can also be attached to a water
pump, it can serve to power a rice drier, and can be fixed
to a small trailer and used as an inexpensive means of
transportation. Because it can be operated by an individual farmer for many more hours per week than either a
small truck, a diesel pump or an ordinary tiller, it is
an extremely economical machine. If different machines
had to be used for each of these functions, problems of
maintenance, supplies of spare parts and servicing would
be more difficult to handle. Clearly, the fact that such
a machine can be used for a wide variety of functions also
means that none of them can be carried out quite as
efficiently as with a more specialised machine. In the
same way, a multi-purpose dam, aimed simultaneously at
flood control, irrigation and electricity generation, may
be less efficient in each of these functions than a more
specialised dam, but in overall terms less expensive, and
relatively more efficient because it can fulfill several
functions.

Some types of technology may be inherently more multi-
purpose than others. But in most cases, as illustrated by
the Southeast Asian tillers, the multiplicity of functions
is something that can be built into a new technology at the
design stage: if such a tiller is to be used also as a
transportation vehicle, it must among others be more powerful than would be required if it were used only for
tilling.

One interesting case of a technology which is multi-purpose
by nature rather than by design is ferro-cement technology
(i.e. cement reinforced with wire mesh, or natural products
such as jute or bamboo). The same basic technology can be
used for making irrigation pipes, water tanks, food storage
vessels, and with a few minor modifications (e.g. in the
proportion of cement to other materials) it can be used as
a roofing material and as panels for house construction.
The major advantage of this multiplicity of applications is
that once a poor farmer, or for that matter a village
craftsmen, is taught the basics about this technology and
about one or two major applications, he can apply this
knowledge to a very wide range of other technical problems.

Clearly the search for multi-purposeness in technology

should not blind us to the inherent advantages of certain
extremely narrow and specialised appropriate technologies.
One example here might be that of very low-cost disposable
syringes that are designed in such a way that they physically cannot be used a second time, thus obviating one of
the major health problems in poor countries seeking to
promote mass vaccinations, namely, the multiple use of disposable syringes that cannot be sterilised after they are
used the first time.

What these few examples suggest is that the concept of
multiplicity of applications needs to be considered more
systematically than it is at present. A multi-purpose
technology is not inherently better than a single-purpose
technology, but this criterion is worth exploring and
might serve as a useful guideline in the design of new
appropriate technologies.

CONCLUSION

The above discussion leads us to conclude that no technology however good, can score high marks in the light of all
the criteria discussed in this paper. Some sort of balance
needs to be found among the potentially relevant criteria.

This paper is not a conclusion in the debate about the
criteria of appropriateness. It should be seen first and
foremost as a set of questions, not as clear-cut answers,
and our aim here has merely been to suggest that the
emerging second generation in appropriate technology will
have to raise questions which until now have been somewhat
neglected, or at least analysed primarily in economic
terms. Economics is of paramount importance, but there are
also a number of other elements which must be considered
more carefully and more systematically.

Chapter 2
TECHNOLOGIES APPROPRIATE FOR A BASIC NEEDS STRATEGY

A.S. Bhalla

INTRODUCTION

In the current debate on national and international aspects of development, three types of questions are increasingly coming to the forefront:

(a) What is the meaning of the basic needs (BN) concept, approach and strategy of development?

(b) What production structure and technology-mix are appropriate for such a strategy? and

(c) How can the implementation of a basic needs strategy be reconciled with the implementation of a New International Economic Order (NIEO)?

The present crisis in the world economy and the inequitable distribution of wealth and resources between nations is also reflected in poverty and squalor within nations, particularly in the developing countries. The implementation of a NIEO in an increasingly interdependent world is possible only if structural and institutional changes at the international level are accompanied by changes in national policies and programmes to eliminate poverty and fulfil basic human needs. The basic needs of the poor of the world cannot be satisfied without reshaping the global

_____

[1]Chief, Technology and Employment Branch, International Labour Office, Geneva.

economic order; in its turn, the new Order would not be
worth aiming at if the poor did not benefit from it.

The concept of basic needs is not new. What is novel however is a systematic focus of attention on the fulfilment
of basic needs, especially of the poor target groups, in a
specified short time-span, as one of the vital objectives
of development plans, policies and programmes. If basic
needs are to be provided for in a shorter time-span than
the conventional growth strategies would permit, emphasis
on the pattern and style of growth becomes much more
crucial than on overall growth per se.

We argue that a basic needs approach is in fact a more
comprehensive view of development, as it calls for a
greater emphasis on indigenous technology generation within
developing countries, a larger capacity of the productive
system to absorb the benefits of science and technology,
and a wider spread of the benefits of growth through decentralised production and consumption planning.

More appropriate technologies are needed by the developing
countries regardless of the type of development strategy
adopted. A basic needs objective, defined as improvement
of access to goods and services through appropriate institutions, adds additional dimensions to the concept of
appropriate technology. Thus, the focus of attention
should be as much on technologies for marketing, distribution and transport as on technologies for production.

This paper is concerned with an inquiry into what consti-
tues a concept and strategy of basic needs; how a basic
needs strategy is different from the conventional strategies of development, what the implications of such a
strategy are for production planning, commodity-mix and
technology choice and development. It starts with an
outline of a basic needs approach and strategy and goes on
to sketch its possible technological content. Special
emphasis is laid here on the implications of national/sub-
regional policies of (collective) self-reliance and the
New International Economic Order (NIEO) for productivity-
raising and employment-generating technical change in the
rural and urban informal sectors. We also consider
examples of the types of technologies the generation and
utilisation of which would be necessary for basic needs
fulfilment. Finally, we examine the crucial questions of

national and international action.

A BASIC NEEDS APPROACH AND STRATEGY

In its simplest form, a basic needs approach can be defined
as a type of development that attaches a special weight to
the fulfilment of basic needs, both material and non-
material, in a given society and aims at meeting this
objective in the shortest time-span. Since the basic needs
of the poorest are less likely to be fulfilled in the normal
course of development, they should be identified as the
priority target group. In this sense, this approach to
development is not fundamentally different from "employment-
oriented" and "anti-poverty" approaches. It also aims at
rapid growth, more equitable income distribution and removal
of poverty, but the paths taken to achieve these goals may
be quite different.

To achieve these objectives the "income" or "poverty"
approach may not necessarily ensure the fulfilment of
basic needs of the bulk of the population. For one thing,
income-generation alone may not ensure minimum private consumption and provision of essential public services. Additional income may be spent on food items of lower nutritional value which may be supplied by mass-production
methods. The income approach has tended to neglect the
need for the supply of appropriate products to ensure that
the purchasing power is actually translated into needs fulfilment. On the other hand, a basic needs approach brings
access to goods and services to the forefront of development objectives. It attaches special importance to finding
out the reasons for the failure of goods and services to
reach the target groups for whom they were intended, and to
the need for ensuring their availability.

Access may be made possible in a number of ways, one of
which is employment-creation. Through the generation of
incomes, employment-creation provides an opportunity for
the poor to have access to education, food, shelter,
clothing, etc. Employment is a direct means of generating
incomes for those who without work would have to depend on
doles or support from extended households. Employment-
creation may also be a better means of redistributing incomes than straight government transfers in the form of
doles. For one thing, the possibility of corruption is

less likely when income is redistributed through employment than through subsidies.

In a somewhat different manner, employment also enters as a labour input necessary for producing goods and services for which demand is created through generation of incomes to the poor. However, the employment and production aspects are difficult to separate in the case of the self-employed whose income may be directly related to the productive activity in which they are engaged. The fact of being employed for a wage or on "own-account" also gives a sense of recognition, pride and participation. Qualitative aspects of employment, that is making work more human and congenial, apart from the monetary compensation it carries with it, may be an important consideration related closely to the sense of participation and recognition.

However, employment-generation as such need not necessarily ensure either access to goods and services or fulfilment of basic needs. Even apart from non-fulfilment of needs for public services, constraints in the supply of private consumption goods such as foodstuffs could lead to inflationary pressures. It is therefore essential that increased incomes to the poor are matched by a corresponding availability of basic wage goods.

It follows that both generation of employment and incomes are necessary but are not a sufficient means of ensuring access to goods and services.

Lack of access is due not only to an inadequate supply of available goods but also, and often, due to faulty distribution system. An emphasis on the supply of goods should therefore involve (a) overcoming the supply constraints by ensuring fuller utilisation of available productive capacity as well as expanding this capacity, (b) improving distribution channels and delivery systems, and (c) avoiding waste and hoarding, etc.

Unequal access to private goods and public services may result also from the prevailing structure of unequal income distribution. Empirical evidence shows that several countries which witnessed rapid growth and equitable distribution, enabling fulfilment of basic needs (e.g. Korea, and Taiwan between 1950 and 1975) started with "relatively equal asset and income distribution; many of those that

experienced rapid, inequitably distributed growth began
with sharply unequal income distribution".[2]

Thus, basic needs fulfilment is a socio-economic and
political problem as well as a technological one. Structural changes and political commitment are essential pre-
requisites of a basic needs strategy. Even in countries
like India and Sri Lanka the minimum needs-based programmes
proposed by the planners in the past were not always fully
implemented. The failure to implement these plans and programmes seems to have been due to the absence of necessary
institutional reforms. Lack of administrative structures
and capacity, required particularly to implement intricate
programmes, tend to prevent fulfilment of basic needs
targets. Traditional types of administrative systems and
structures are often inadequate to implement decentralised
programmes in rural areas requiring local knowledge,
cultural identity, and a sense of motivation. They are
equally inappropriate for ensuring the delivery of public
services to the target groups.

Both poverty-oriented "income" strategy and "BN" strategy
are aimed at removing inequalities within given societies.
These inequalities may range from inegalitarian income distribution to the non-economic inequalities of status, such
as unequal participation in decision making, freedom of
choice, authority and access to it, and satisfaction from
work and recognition from employment. While a poverty-
oriented strategy is confined almost exclusively to the
alleviation of economic inequalities, a basic needs strategy
also encompasses elimination of at least some of the non-
economic inequalities. The latter is a necessary, but perhaps not a sufficient, condition for the satisfaction of the
basic needs of the masses in the shortest possible time.

One way to alleviate inequalities especially of the non-
economic type is to ensure wider participation of the
potential beneficiaries of development in planning and de-

---

[2]See David Morawetz: "Twenty-five years of economic
development", Finance and Development (Washington, DC.),
Sep. 1977. Also see Chenery, et. al., Redistribution with
Growth, Oxford University Press, 1974.

cision making. Without popular participation the perception of development is likely to be lopsided. Unless
people specify their needs (through a process of collective
decision making or otherwise) any targets for supplying
goods and services set by the central planners are at best
likely to be either off the mark or geared to the needs of
the social groups who are already much above the minimum
level of poverty.

Where direct participation is not feasible in practice,
articulation of poor people's needs can be ensured through
such institutions as cooperatives, peasant organisations,
communes, trade unions and similar associations.[3]

Popular participation enables decentralisation of decision
making in planning and production. It can take several
forms: for instance, it may imply that people themselves
determine their basic needs. Secondly, it may mean that
people participate in setting local targets for production
and consumption, and ensure their implementation. Decentralisation in decision making facilitates autonomy at
the local level. In most developing countries, one of the
major constraints to the implementation of development programmes is an inadequate institutional and organisational
structure. Under a centralised system, transmission of a
decision from the centre to the periphery tends to take
time and is likely to get distorted somewhere along the
line. On the other hand, decentralisation is likely to
help overcome unnecessary bureaucratic and administrative
delays.

Basic needs planning requires participation, decentralisation and local resource mobilisation combined optimally
with a dose of central coordination and planning. The degree of local initiative and participation, the effectiveness of social and economic institutions, and the quantity

---

[3]This, however, presupposes that these institutions are
effective in representing the interests of the poor, that
they are in regular and close contact with the larger
poverty groups in the rural and urban informal sectors, and
that they are useful instruments in organising the poor
into an articulate political force.

and quality of indigenous material and human resources,
will largely determine the nature of technology which in
turn will affect the cost of provision of basic needs.

Fulfilment of basic needs cannot be the sole responsibility
of national action by the developing countries. While it
is the sovereign right of each developing country to define
its own basic needs objectives and the means to fulfil them,
the inclusion of basic needs as a part and parcel of a more
equitable international (economic) order would involve reshaping the aid and trade policies of the advanced countries,
redefining the role of private foreign investment and of
multinational companies and controlling the flow of the most
advanced technologies from the North to the South. Reorientation of these international measures is essential in
order to make the efforts of the developing and developed
countries compatible in launching an attack on world poverty.
There is however no substitute for the full political commitment by the developing countries themselves to the implementation of appropriate reforms and institutions to satisfy
the basic needs of the poorest in particular, and of the
population in general.

The implementation of basic needs by the developing countries
may also require a more self-reliant development implying
less dependence on the advanced countries. Self-reliance,
both national and collective (interpreted in terms of cooperation among developing countries) is a prerequisite for
improving the bargaining position of the developing vis-à-
vis the developed countries. There is a parallel here with
the organisation of the rural poor in a developing country.
In the same way as the poor need to be organised into
associations to raise their political power, the developing
countries need to organise themselves into groups to overcome their weak bargaining position and to safeguard themselves against economic disturbances like recession and
inflation in the advanced countries.

TECHNOLOGICAL CONTENT OF A "BN" STRATEGY

Be it the distribution of income and purchasing power, the
composition of output, access to resources, inputs and
outputs, all these variables have important implications
for the choice and development of technologies in the developing countries. This paper does not claim to present

a technological blueprint for a basic needs strategy. It
is a more modest attempt to outline technological implications of the following elements of a basic needs strategy:

(a) incomes to the poorest target groups;

(b) access to goods and services and inputs;

(c) physical production of goods and services;

(d) participation and decentralised production; and

(e) national and collective self-reliance.

We shall consider below the relevance of each of these
components for technology.

Incomes to the Poorest

Generation of income to the poorest is possible in a number
of ways, e.g. by employment, through income transfers by
the government, through redistribution of assets such as
land, and through a rise in the productivity of existing
methods of production. The traditional techniques are no
doubt employment generating but their productivity is
often so low that they may be "inappropriate" for generating adequate incomes. In fact the technologies actually
in use in the small and urban informal sectors may be too
labour-intensive. Their adaptation through technical
progress within these sectors may call for a somewhat
higher capital to labour ratio, thus implying less employment intensity within the process itself. Thus, technological innovations for removing drudgery of women may well
be labour-saving but income-generating, and hence desirable.

In the past, too much emphasis has been placed on international technology transfers and too little on technology
gaps within developing countries. This has perpetuated
the technological vacuum in the traditional sectors and
has tended to widen the gap between the traditional and
the modern sectors.

A basic needs strategy should aim at raising incomes of the
poor through the development and improvement of traditional
technologies in the rural and urban informal sectors. For

one thing, it may often be easier to modify and adapt
existing technologies than to create them de novo.
Secondly, the problem of accessibility to the poor is
minimised when attempts are made to improve on the technologies to which they are already accustomed, but w .ch
without modifications might not generate reasonable levels
of productivity and income. An ILO/UNDP project j
Tanzania demonstrated that in a subsistence economy prevailing in the region considered, the initial cash outlays
required for tractorisation or even for implements considered "intermediate" were well in excess of what the poor
farmers could afford. The project therefore recommended
that emphasis should be placed, at least initially, on the
design and development of those village technologies which
are more productive than the traditional ones, yet within
the reach of small farmers and other poorest groups of the
population.[4]

There is as yet some controversy regarding the extent to
which it is possible to upgrade existing technologies
through modifications and adaptation. There are those who
argue that traditional technologies, even though inefficient (in terms of low productivity), may be appropriate in the context of the rural sectors on other counts,
viz. the local needs and the skill patterns, not to speak
of the socio-cultural environment. They believe that it is
preferable to introduce improvements of these technologies
through the application of "modern" scientific and technological know-how available within the so-called "modern"
sectors of developing countries. Recent attempts to improve the bullock-carts and earth-moving methods (wheelbarrows, wooden stretchers, etc.) are but a few examples
of the scope for upgrading the existing technologies.
There is another school which would discard existing "traditional" technologies and argue for the use of "modern"
technologies to enable a quantum jump in the standards of
living of the population. Whether certain existing technologies can be adapted/improved or not is an empirical
question which needs to be treated case by case.

---

[4]See George McPherson and Dudley Jackson, "Village
technology for rural development", International Labour
Review, February, 1975.

Concentration on improvement of technologies alone as a
means of income generation will at best be an insufficient
solution. Redistribution of investment resources in favour
of the hitherto neglected urban informal and rural sectors
would be necessary at the same time to ensure that the
small-scale producers have wherewithal to effect necessary
technological improvements. This may call for redistribution of assets like capital and land to the poorest groups
of society, along with a necessary public provision for R
and D work to respond to their production and consumption
needs.

Access

The capacity of the small farmers and small producers to
innovate very much depends on their access to various inputs. The experience of the Green Revolution has shown
that in many cases the larger farmers benefited much more
from the use of HYVs of rice and wheat than the small
farmers. This does not imply that the new technology was
inappropriate; rather the institutional arrangements for
credit, land tenure and market structure were biased in
favour of the landlords and the big farmers. The experience of Asian countries, notably India, Pakistan and
the Philippines, shows that the small farmers had little
access to such inputs as fertiliser, water, credit and
technical knowledge. On the other hand, these inputs were
made available to large landowners on subsidised rates to
ensure that they innovated in order to raise agricultural
production rapidly.

The experience of agriculture also applies to the industrial
sector. Market imperfections, that is, factor price distortions in labour and capital markets, result in the relatively low cost of capital and a relatively high cost of
labour facing the modern manufacturing sector, and a high
cost of capital with a relatively low cost of labour in the
"non-modern" sectors. Work undertaken on the urban informal sector under the ILO World Employment Programme
(WEP) suggests that the informal sector enterprises either
do not have any access to credit at all, or they have a
very limited access characterised by exorbitant interest
rates and stringent terms and conditions. In the case of
Ghana automotive repairs, only 6 per cent of the masters
borrowed the initial amount necessary to start busi-

ness.[5] Credit was the least important source of financing
since the small enterprises had little access to credit
institutions. One of the sources of financing was the
masters' savings out of wages earned in the modern sector
during the course of apprenticeship training there.
Another was the customers. In the case of auto repairs,
customers trust masters and make cash payments before
repairs are done. This enables the masters to buy the
necessary raw materials. Such pre-financing works to the
mutual advantage of the repair shops and the customers.
The shops obtain necessary materials without having to
keep big amounts of inventory stocks: the customers are
assured of low repair costs, at least lower than what they
would have to pay to the large modern sector workshops.

Access to inputs can also be related to the access to
product markets. The informal enterprises may be forced
to sell goods to a single buyer or a small group of
buyers on whom they have to depend for credit and other
needs. Access to product markets can be further reduced
if the small-scale enterprises have to compete with larger
modern enterprises in the same market. However, this need
not be the case when the "informal" and "modern" sector
products are non-competing. A case study of tin cans in
Thailand[6] represents an interesting example of how sea-
sonality of demand and small segmented markets worked in
favour of small producers with low capital-intensity.
Local demand for a variety of canned goods like fruit,
vegetables and pickles led to the emergence of small-scale
plants owned by local businessmen. In the early 1960s
large foreign firms entered Thailand to produce canned
dairy products and canned pineapples for the foreign
market. The rapid growth of the "foreign" sector producing canned goods led a specialist can-making firm to
enter the market in the late 1960s. Thus, a dualistic

---

[5]A. Hakam, Technology Diffusion from the Formal to the
Informal Sector: Automotive Trade in Ghana, WEP Working
Paper (WEP 2-22/WP.35), ILO, Geneva, July 1978.

[6]R. Bell, et. al, Industrial Technology and Employment
Opportunity, ILO, Geneva, 1976 (mimeo).

structure of can-making industry has emerged: with the informal segment characterised by low product quality, low income local markets, lower wages, limited access to credit, and use of labour-intensive "semi-manual" techniques. The modern sector caters largely to the high income domestic and foreign markets. The large firms have not entered the markets of the informal producers partly because a sizeable proportion of the production of these firms is for their own use. Secondly, the pattern of demand in the informal sector requires production of a different kind: a larger number of different shapes and sizes of open-top cans to which the automated high-speed equipment of the large-scale modern sector is not geared.

The question of access is also in some ways related to information and knowledge about the efficient technologies developed either within the rural and urban informal sectors, or those available and in use in the modern sector. This raises questions about the need for linkages with producers in the rural and urban informal sectors through extension services and other means of diffusing new and adapted technologies. Agricultural extension services have already been well established in many developing countries. There is an equally urgent need for industrial extension services to reach the small informal sector enterprises in order to assist them in product and process adaptations and in the use of new technology. Another method of raising the technological levels in these enterprises is to improve their linkages with the modern sector through subcontracting which makes technological improvements among subcontractors possible. Watanabe[7] has distinguished between three types of linkages:

(a) input linkages which refer to supply of raw materials and machinery. Such linkages may for instance take place between petty garment producers and suppliers of textile fabrics or sewing machines;

---

[7] See S. Watanabe, Technological Linkages between Formal and Informal Sectors of Manufacturing Industries, WEP Working Paper (WEP 2-22/WP. 34), ILO, Geneva, March 1978.

(b) <u>market linkages</u> between parent firm and subcontractors which, unlike the above, supply the
same end product to the same customers; and

(c) <u>technological linkages</u> which imply flows of
technology and skills between the two sectors.

Not enough is yet known on the precise form and the extent
to which linkages exist between formal and informal sectors,
on the manner in which entrepreneurs and workers in the informal sector acquire skills, the extent to which such
skills are adequate for innovations in the sector, on ways
to encourage the transfer of technology from the formal to
the informal sector, and on whether such transfers are
necessarily beneficial to the development of the latter.

The experience of Japan suggests that rapid industrialisation was made possible to a large extent by close collaboration between the large-scale and small-scale sectors,
with the former helping the latter to upgrade and adapt
their technology. In the case of China, it is reported
that, at the earlier stages of development, linkages were
mainly one-sided, with modern industry providing the rural
industry with the necessary technology. When rural industry developed, these linkages became mutual, and in the
third stage of industrialisation the two sectors became
more or less integrated. The improvement in the quality
standards in the rural sector permitted an expansion of the
subcontracting system.[8] These are but two examples of the
positive aspects of inter-sectoral technology transfers.
There may however be negative factors at work too. The
flow of technology from the modern to the traditional and
informal sectors may actually displace the latter technology, thus resulting in considerable unemployment.
Secondly, the modern sector technology may be too advanced
and hence inappropriate for the rural and urban informal
sectors. Thus, while improvement of technologies in the
urban and rural informal sectors should be an important aim
in a basic needs strategy, this goal could perhaps be
better achieved by local innovations within these sectors
through public R and D support, and through access to

---

[8]See Jon Sigurdson, <u>Rural Industrialisation in China</u>,
Harvard University Press, 1977.

finance capital.

Another aspect of access with a technological dimension is
the provision of access to public goods and services. Be
these inputs or outputs, access is often prevented due to
the lack of adequate delivery systems to channel such
services to the urban and the rural poor. Improvement in
delivery of government services calls for an investigation
of the desired pattern of output and the infrastructure requirements, the use of technology and the cost of making
these services and infrastructure available to the poor
under different organisational and technological forms.
For example, the provision of safe water for all, recommended at the World Employment Conference, is one of the
goals of a basic needs strategy. A large majority of people
in developing countries do not have an easy access to safe
water supply. A major constraint to the development of
adequate water supplies and water treatment facilities is
the shortage of funds. There is therefore a need to develop
and implement technologies which minimise the cost of water
treatment and water distribution systems. This is particularly true for rural areas where diseconomies of scale
raise production cost of treated water per unit of output,
and where people may not be able to afford treated water
unless it is subsidised by the government.

In the final analysis, access to inputs is tied up with
asset redistribution and payments system (i.e. who determines who gets what sort of work and with what tools/equip-
ment/land). Thus, the modern sector's privileged access to
capital and other inputs need not necessarily be due to
factor price distortions which are often believed to cause
the use of inappropriate technologies. It may well be the
institutional factors and unequal pattern of income distribution that determine factor prices much more than
market forces.

Physical Production of Goods and Services
──

The question of ensuring access is not very meaningful
without a production structure that has the capacity to
ensure the availability of those items of private and
social consumption which are essential for implementing
basic needs targets.

A basic needs strategy implies a production pattern and
a product-mix different from the one under a conventional
growth-oriented strategy. The objective of satisfying
private and social consumption requirements of the bottom
20-30 per cent of the population requires a very different
consumption bundle of goods and services from the one that
would obtain if equal weight were given to the requirements
of the rich and the poor.

There are those who argue that the pattern of consumption
and of industrialisation (i.e. the type of product-mix)
largely determines the choice of technology. Implicit in
this argument is the assumption that very few goods can be
produced with more than one technique of production. At
least in certain sectors technical choice is possible even
without changing the composition of output. Since the basic
needs strategy gives higher priority to the production of
goods and services to be consumed by low-income groups, it
requires the development of technologies which can produce
simple goods with simple "characteristics" at low cost.

Basic needs targets need to be established in physical
quantities of goods and services so that shortfalls in
their fulfilment can be measured regularly. Goods and
services may be defined in the form of their principal
characteristics, e.g. calories, cubic metres of space or
metres of cloth. Although the basic human needs for food,
shelter and clothing are universal, the ways and the extent
to which these needs are fulfilled will depend, by and
large, on the physical availability, characteristics and
distribution of these goods. In a basic needs strategy,
goods may be considered "inappropriate" if they represent
"characteristics" which are beyond the reach of the poor
target groups, or are not wanted by them.

A basic needs strategy will require more emphasis on product development and adaptation than is generally placed
on these aspects of technology in production planning.
Since products are indivisible in the sense that the purchase of a given product implies the acquisition of a number
of characteristics simultaneously, production planners may
have to allow for product substitution: choice in favour of
products which do not have excessive characteristics or
standards. Standards and characteristics could be defined
as "excessive" if they were beyond the reach of the consumers below a certain level of income, or if they were

unwanted by a set of consumers whose basic needs bundle of
consumption goods does not include these characteristics.

Just as the basic needs targets are not determined for all
times but need a constant revision to take account of
change in incomes and living standards, so do the
characteristics change with rising material standards.
Therefore what are "appropriate" standards today may no
longer be appropriate a few years hence. It is necessary
that like basic needs targets the product standards and
characteristics be revised periodically.

Since individuals demand products with a number of
characterisitics, a technology component of a basic needs
plan can be described and analysed in terms of product
innovations referring to a particular bundle of goods
required for the fulfilment of basic needs. These innovations may take the form of development of new products, or
adaptation of existing ones. A major effort would be required to (a) adapt end products to lower or different
specifications, simpler packaging, modelling, etc., and
(b) improvement of quality of traditional and unsophisti-
cated products. The end purpose is to ensure that unnecessary characteristics are dispensed with, and products
are made to require less maintenance. Similarly, greater
emphasis will need to be placed on "utility" versions of
modern sector goods, and a reduction in the production of
"luxury" goods. A number of foreign car manufacturers
established in the Philippines have introduced "Asian car"
models "in order to increase the local content of production without raising costs or sacrificing quality to any
appreciable extent. An extreme case is provided by one
manufacturer in whose car bodies bamboo and wood, which
are in plentiful supply in the Philippines, are used".[9]

*   *   *

Having considered the three principal characteristics relating directly to technology choice and development, we

---

[9]ILO, <u>Technologies for Basic Needs</u>, by Hans Singer,
Geneva, 1977, p. 54.

now examine two additional features of the strategy, namely (a) popular participation in decision making, and (b) collective self-reliance and technical co-operation among developing countries (TCDC). Both these features have implications for the organisation of production and its distribution.

Participation and Decentralised Production

Popular participation in a basic needs strategy essentially boils down to decentralisation of production and to planning from below. Production and consumption planning can be either centralised or decentralised, or a combination of the two. With perhaps a few exceptions (e.g. Sri Lanka and China) most planning in developing countries is highly centralised. Investment allocation, production and consumption targets, balances between demand and supply, are all considered by national planners on the basis of certain assumptions about aspirations of the population within the framework of declared goals of planning.

A distinction needs to be made between planning and decision making. Many countries have centralised plans but effective decisions are taken at the local level, while in others the reverse may apply, e.g. where large state corporations or multinational companies make important decisions.

In any developing society, there are a number of diverse decision-makers ranging from large-scale producers including national and multinational enterprises, national governments (who in certain cases also participate directly in the production process through state-owned corporations and public enterprises), to self-employed individuals in family enterprises, co-operatives or extended households in the urban informal and rural sectors. All these groups influence technology choice and development through their participation in economic activity. One may therefore ask the question: what is new, that a basic needs strategy really calls for? A move towards this strategy implies that:

(a) the weight of small-scale producers at the grass-roots level increases;

(b) those who do not directly participate in a productive
economic activity also have a say in matters that influence their welfare;

(c) scope for self-employment and household modes of production where ownership and decision-making are combined
should be increased; and

(d) the rural poor who form the bulk of the population are
organised in order to carry a political weight which
will enable them to articulate their needs.

In a conventional development strategy, resource allocation
is skewed like the income distribution with the result that
the national resources and assets are controlled by small
groups of powerful vested interests. These resources are
therefore used largely to satisfy the demand pattern which
serves their interests, leaving the basic needs of the bulk
of the population unfulfilled, both in terms of techniques
(confining employment to few) and products. A shift towards
a basic needs strategy makes it necessary to ensure not only
greater access to resources but also greater control over
them. These are in fact the prerequisites for the development and application of technologies appropriate at the
grass-roots level. The centralised nature and structure of
social institutions and patterns of ownership can often inhibit the implementation of technological alternatives that
are known and may be socially as well as technically desirable.

In principle, the cooperative forms of organisation and
the framework of household production may be much more
advantageous for the utilisation of socially desirable
employment-creating technologies. This is so because work
is done largely outside the wage system, so that reward for
work is in the form of greater output rather than wage-
payment. The cooperative and household modes of production
have another advantage over the wage system, namely that
workers do not feel alienated. Organisation of production
in a large number of small enterprises in a household
framework is much more likely to ensure that the income
generated is appropriated by family or cooperative members
than by the large-scale producers under a more centralised
production system.

Decentralised production also helps promote learning-by-

doing types of innovations outside the formal R and D
institutions. In the past, in most developing countries
appropriate technologies have been neglected both within
and outside the formal R and D system, partly because of
a heavy dependence on developments in the advanced
countries. Imported modern technology, by destroying the
traditional small-scale productive activity, often pre-
vents learning-by-doing by the semi-skilled and skilled
artisans. It also tends to orient the educated and
scientific elites of the developing countries towards the
problems and standards of the rich countries. The result
is a sense of inferiority and a tendency to imitate rather
than innovate. The encouragement of decentralised production at the local level should help overcome these
negative tendencies.

The experience of Sri Lanka, with the District Development
Councils (DDCs), and more recent attempts of the Indian
Government at block-level planning, both reflect a realisation that the development effort at local level alone can
be successful in estimating people's needs and aspirations,
as well as the availability of resources and their mobilisation. The DDCs in Sri Lanka represented one type of institutional framework that enabled local population to
participate in political, social and economic activity, and
to rely, in so far as possible, on local skills, locally
developed machinery and technology. What was initially a
purely technical/administrative unit was subsequently re-
organised into a politically based programme with the
appointment of a local government nominee or a member of
parliament as the chairman of the DDC. This politicisation
seems to have been based on the recognition that a political
commitment and support at the local level is essential for
the implementation of policies and programmes.

Decentralised production and technological development do
not imply elimination of planning at the national level,
although too much detailed planning might not be necessary
if the right kinds of policies are introduced. What the
basic needs strategy implies is an aggregation of local,
district and provincial plans which reflect more realistic-
ally the aspirations and needs of the people for whose
benefit planning is designed. Such an aggregation helps
the central planners to cross-check their own notions of
what the local resource availabilities and skills are for
producing the target product-mix with the appropriate tech-

nology-mix.

An effective mix of central and local planning needs to be combined with adequate and timely plan implementation. An efficient administrative system is a prerequisite to the successful implementation of any plans or policies. A development strategy based on decentralisation and dispersal of economic activity tends to make much greater demands on the administrative structure and capacity in the developing countries. An adequate administrative structure for a basic needs strategy will require appropriate local institutions at the district and block levels. It would also require adequate communication system whereby information and decision signals can be exchanged promptly between the national level and the province, district or county. Successful development and diffusion of technologies through decentralised organisation presupposes a proper coordination and dovetailing of production and investment plans at different administrative and economic levels.

National and Collective Self-Reliance

Self-reliance at the national level may imply lesser dependence on the advanced countries for import of technologies and products. It does not however mean total autarky or independence. Collective self-reliance and technical cooperation among developing countries are simply an extension of "local" and "national" self-reliance to a sub-regional and regional level. It provides for the exchange of technologies and goods among developing countries without however excluding trade and technology transfer between developed and developing countries as equal partners. It also implies generation and adaptation of indigenous technology with a view to developing a strong national technological capability within developing countries. Weak science and technology systems in developing countries make them dependent on the advanced countries in two principal ways: (a) they are forced to use advanced technologies from the developed countries regardless of their appropriateness, and (b) the weak bargaining and negotiating position of developing countries either reduces their access to modern technology or raises the cost of technology transfer.

Collective self-reliance can be defined in terms of a reduction of "technological dependence" of the developing

countries on the advanced countries whose problems and resource endowments are very different. Technological
dependence may be disadvantageous and harmful to the
objectives of a basic needs strategy in several ways.
First of all, it reduces control over technological decision-making and thus conflicts with the objective of
participation at all levels. In a study of totally local-
owned firms in Argentina which buy their technology abroad,
Sercovich concluded that the nature and "content of technology agreements is such that most of the power of independent decision-making is taken out of the hands of the
local owners and managers".[10] Secondly, it may hinder the
indigenous development of scientific and innovative
capacity.[11] Free transfer of technology from the advanced
countries tends to reduce the motivation of the developing
countries to develop their indigenous technology. Even if
they have such motivations, they cannot withstand the
dominance and competition of imported technology without
the "infant-industry" protection to local technology.
Further, contracts governing the transfer of technology
also inhibit local R and D to develop indigenous technology
and adapt imported technology.

Promotion of technological independence may imply redirec-
tion of existing North-South links. It may also require
greater national self-reliance. There is controversy on
whether cutting off ties with the advanced countries is
beneficial to LDCs in respect of a self-reliant technological development. The experience of countries like China
suggests that the discontinuation of Russian technical
assistance and supply of equipment in the fifties stimulated
local technology development during the Great Leap Forward

---

[10]F.C. Sercovich, "Foreign technology and control in the
Argentinian industry" (Ph.D. thesis, Sussex University,
1974) quoted in Frances Stewart, Technology and Underdevelopment, 1977, p. 131.

[11]See A.S. Bhalla, "Transfert de technologie, technologie
appropriée et emploi", Revue Tiers-Monde (Paris),
janvier-mars 1976.

period.[12] This was perhaps made possible by the existence
of local capacity to innovate.

National independence combined with "horizontal" cooperation among LDCs is likely to be much more conducive
to the building of national technological capability and
development of technologies appropriate for the socio-
economic conditions of the developing countries. Such an
exchange between Third World countries would also reduce
the costs of technology transfer by raising their bargaining position. Furthermore, in the case of small-sized
countries, the national units of R and D may be too small
to be economically viable; sub-regional R and D among a
group of LDCs may promote economies of scale.

TCDC in the field of technology can take many forms.
First, flows of technological information among developing
countries need to be improved through exchange of technical
personnel and know-how. Such information may cover
available technological alternatives, sources of technology
supply, terms and conditions of their transfer, technology
policies, and existing research and development facilities.

No institutional machinery at present seems to exist for
specialised exchange among LDCs. However, the sub-regional
groupings like the ASEAN and other bodies of the developing
countries like the Group of 77 and the Non-Aligned Group
provide a rudimentary beginning towards collective self-
reliance. While it is true that until recently the LDCs
did not make a political commitment to appropriate technology, this position has witnessed a welcome change.
Countries like India and Sri Lanka have declared appropriate
technology development as their official policy. The ASEAN
have put appropriate technology on the agenda of their
ministerial meetings on a regular basis. At the ASEAN
Labour Ministers meetings in Baguio (Philippines, 17-19 May
1976) and Pattaya (Thailand, 24-25 May 1977), the participating countries agreed to co-operate with each other in
promoting the development and transfer of appropriate

---

[12] See A.S. Bhalla, "Technological choice in construction
in two Asian countries: China and India", World Development, March, 1974.

technology. They underlined the need for governments "to
establish a national organisation charged with promoting
the use of appropriate technology throughout the decision-
making apparatus of the public sectors. This organisation
could have its representatives in each government ministry
who would assess every opportunity in the ministry for
introducing appropriate technology at each decision
level".[13]

The Group of Non-Aligned Countries has also been preoccupied with action needed for the application of
appropriate technology in developing countries. At their
Fifth Summit Conference in Colombo in August 1976, the
Group elaborated an "Action Programme for Economic Cooperation" of which appropriate technological development
formed a prominent part. At their meeting in New Delhi
(7-11 April 1977)[14] the Bureau of the Non-Aligned Group of
Countries decided to set up an Intergovernmental Working
Group on the Applications of Appropriate Technology.

More recently, the First Conference of Labour Ministers
of Non-Aligned and Other Developing Countries (Tunis,
24-26 April 1978) drafted the Tunis Action Programme which
contains a number of activities for further horizontal cooperation in the fields of employment, training and education, and appropriate technologies.[15]

---

[13]Report of the Special Meeting of ASEAN Labour Ministers,
Baguio City, Philippines, 17-19 May 1976, pp. 154-155.

[14]Report of the Bureau on the Meeting of the Non-Aligned
Group of Countries, New Delhi, 7-11 April 1977 (NAC.5/BUR.1/
DOC.2/REV.2).

[15]Action Programme of the Tunis Conference - First Conference
of Labour Ministers of Non-aligned and Other Developing
Countries, Tunis, 24-26 April 1978.

TECHNOLOGIES APPROPRIATE FOR BASIC NEEDS SATISFACTION

It follows from the foregoing that appropriate technology is a concept which implies a suitable policy framework and a broad-based pattern of development. It is not a tool-kit of a selected number of technologies with technical and engineering specifications which could fit in with the requirements of all Third World countries. Peculiarities of local conditions (economic, social, ecological, etc.), differences in socio-economic systems and institutions, and specific production and consumption priorities, all these factors rule out a "standard tool-kit" approach. However, although it is wrong to look for particular technologies as appropriate for all climes and times, it is possible to examine broad priority areas in which technologies should be developed and upgraded in a basic needs-oriented strategy. The following are some of the examples of the types of technologies that would require special emphasis:

A.   Pre-harvest technologies

(a) tools and equipment for ground preparation, planting, weeding and harvesting;

(b) water supply and irrigation technologies, e.g. equipment for storing, lifting and moving water.

B.   Post-harvest technologies

(a) food-processing and agro-based technologies;

(b) food preservation, techniques of packaging, etc.;

(c) food storage technologies;

(d) low-cost transport modes in rural areas for marketing of agricultural produce.

Although precise definition of basic needs and the technologies required to provide them will vary from country
to country, depending on its over-all production and consumption priorities, food will perhaps be one of the most
prominent basic needs in most if not all developing
countries. Therefore, research and development on technologies appropriate for food production, its marketing
and distribution, and its processing and preservation,
deserve priority.

Among the pre-harvest technologies, tools and equipment
for crop cultivation and irrigation are perhaps the most
significant topics of research and design engineering.
At present, what is readily available on-the-shelf is
either the most modern and sophisticated agricultural
machinery (big tractors, combine harvesters and the like)
or the most primitive hand tools with very limited life
and low productivity. There is a virtual absence of
"intermediate" technologies with productivities higher than
the traditional tools, yet within potential reach of the
small and the medium farmer. While rapid innovations have
been witnessed in the biological technology, in the form
of HYVs, appropriate technical innovations on the mechanical
side have lagged behind.[16]

Little is at present being done to design equipment for
upland cultivation or animal-drawn equipment. These
small-scale farm tools/equipment deserve much greater
attention in a basic needs strategy: they can be designed
for local manufacture with economical labour-intensive
methods. Take the example of recent attempts at the
design of simple hand-pumps for lifting water for irrigation. A hand-pump is much cheaper than the currently
available diesel pump which apart from its expensiveness
also raises problems of maintenance due to shortage of

_____

[16]There are scattered examples of development and modifications of agricultural implements. The International Rice
Research Institute (IRRI) has designed low-cost tillers,
threshers and dryers more suited to small-scale rice-
farming in South-East Asia.

spare parts. Manually operated shallow tube wells for
irrigation (MOSTI), though a relatively new type of
irrigation technology, are already being used in
Bangladesh.

However, appropriate innovations are required for raising
productivity of technologies currently used for the design
and construction of small-scale irrigation works. Also,
use of appropriate materials is necessary for lining
canals, open wells, and low-cost tube wells.

Much hardware required to upgrade traditional technologies
can in principle be manufactured locally in small-scale
engineering workshops. For example, various types of
devices for lifting water from streams and wells, and
several agricultural tools and equipment are known to be
produced locally. However, their manufacture is not
always economically and technically efficient. The major
obstacle to their local manufacture stems from deficiencies
and inadequate supplies of materials of uniform quality
rather than from any lack of know-how. There is therefore
a greater need for research on materials engineering and
metal-working technology.

We have argued that upgrading of traditional labour-
intensive technologies should be a cornerstone of a basic
needs strategy. A massive technological transformation
that this involves clearly requires machine-fabrication
facilities and well-established metal-working industry.
Supply of adequate materials and inputs such as iron and
steel, coal, and fertilisers, would be desirable. What
is even more important is their appropriate distribution
to the small-scale producers at reasonable prices. There
is thus a need for adequate linkages between large-scale
modern and the small-scale informal sectors. A basic
needs strategy does not therefore imply a complete shift
to the small-scale sector; instead, it underlines the
need for a better integration of different sectors of the
economy so that the productivity and technology gaps between different sectors are reduced.

Under a basic needs strategy capital goods sector has an
important role to play. Machine tools and other capital
goods are desirable not only for rapid accumulation but
also for improvement in indigenous technology and adaptations of imported technology. These industries, in

particular, the machine tools manufacture (which incidentally is more labour-intensive than is often assumed) are the
main sources of technical innovations.

Externalities through learning and diffusion of innovations
to different sectors should facilitate bridging of the
technology gaps between the large-scale and the small-scale,
and between modern and non-modern sectors.

Research also needs to concentrate on the ways in which
technologies can be developed to lower capital-labour and
energy-labour ratios without appreciably raising unit
costs. This raises the related question about the extent
to which "down-scaling" of products and processes is
possible. Experience to date shows that food-processing
industries/technologies can in principle, operate economically at small-scale, although many innovations and adaptations have not yet been applied. For example, in Thailand,
need has been felt for small-scale rice-stabilising units.
Currently, the edible oil contained in bran obtained from
milling of rice degenerates due to delays in its shipment.
This degeneration can be prevented by stabilisation of bran
prior to oil extraction, in small-scale plants located in
rice-growing areas.[17]

Differences in product markets (due to differences in
scale, product characteristics, consumer tastes and product
standards) together with decentralised production structures,
can open up possibilities for the use of relatively labour-
intensive techniques which need to be further explored
within the framework of a basic needs strategy.

One of the reasons for lack of access to food, processed
as well as unprocessed, is the absence of adequate preservation and storage facilities. The perishability of
many food items imposes a need to explore appropriate
techniques for preservation and storage which would enable
food preservation over time, that is, from periods of
abundance to periods of scarcity. The scope for longer

---

[17]UNIDO is currently developing a small-scale rice-
stabilising unit for use in Thailand.

preservation thus provided can enhance the value of food
and food products in terms of the satisfaction of basic
needs.

Appropriate canning and packaging techniques would be
desirable to prevent fresh food from perishing, and thus
to raise food supplies for minimum nutritional standards.
The historical experience of the United States with
"home-canning" of local produce suggests that new decentralised production systems can be combined with the
use of simple relatively labour-intensive techniques to
promote preservation and packaging of food.[18] An interesting feature of the home-canning production system is the
re-use of cans. In rural situations of developing
countries, where location of use of cans and of canned
goods is identical, this re-use is likely to reduce costs
by intensifying the use of material. Depending on the can
costs, canning (for preservation) could be done at the
household level or in small communal "centres" serving a
group of households.

Techniques of foodgrains storage is another area of research that deserves a high priority. Significant quantities of foodgrains are lost in the developing countries
before they ever come to be processed for consumption. A
number of factors account for this wastage: poor storage
and transport facilities, lack of organised marketing
effort, poor management of supply and distribution, and
lack of information on prices and quality of product on
the part of farmers and government officials.

The appropriate choice of technology in grain storage itself on a wide scale can generate both direct and indirect
employment. Improvements in the technique of small-scale
storage by farmers near the point of cultivation, and
choice at the regional and national level between bulk
silos and bagged floor (warehouse) storage for grain depots

---

[18]For an interesting analysis of this subject, see C. Cooper
and M. Bell: Industrial Technology and Employment Opportunity, ILO, Geneva, 1976 (manuscript), Appendix 2 on Product
Markets, Structure of Production Systems and the Choice of
Techniques - A Speculative Case Study.

are the two types of choices, for example.[19]

The next stage after storage is transport and marketing.
Appropriate modes of transport in rural areas are required
to enable speedy shipment and delivery of wage goods and
social services. The goods transport requirements of
farmers cover on-farm transport of seeds, fertilisers and
building materials, etc., and marketing of produce from
the point of collection at the farm to local market or
roadside. The cost of transport modes must be low so that
small and middle farmers can afford to buy or rent them.
It has been shown that for short haul the cost of such
primitive modes as donkeys, wheelbarrows and headloading
is relatively quite high in view of their extremely low
productivity (slow speed). On the other hand, ox and
donkey carts and pedal-powered vehicles (commonly in use
in parts of Asia) offer a potentially low-cost transport
to the small landholder. Nevertheless, very little attention is at present being paid to the improvement of designs
and speed of such transport modes as bullock carts. Increasing emphasis has been placed on low-cost rural access
roads to the neglect of the type of vehicles best suited
for these roads.[20] Absence of low-cost rural transport in
the LDCs forces farmers to rely on such expensive and inappropriate means of transport as headloading. There is an
urgent need therefore to promote R and D work on the improvement of design and manufacture of such vehicles as
handcarts, animal-drawn carts, bicycles and tricycles.

---

[19]See ILO, Appropriate Technology for Employment Creation
in the Food-Processing and Drink Industries of Developing
Countries, Geneva, 1978.

[20]See J.D.G.F. Howe, "Some thoughts on intermediate technology and rural transport", ODI Review, No. 1, 1977; and
IBRD, Appropriate Technology in Rural Development: Vehicles
Designed for On-Farm and Off-Farm Operations, April 1978
(mimeo).

A SUITABLE POLICY FRAME

Policies and programmes of action to apply technologies
appropriate for a basic needs strategy can be considered
at three different levels: local (or micro), national (or
macro), and regional or international. While at the limit
action is to take place at the national level, there are
often factors and decisions beyond the direct control of
the LDCs themselves. Nature of aid and types of trade
(access or lack of it to the markets of DCs) are international factors which can have direct bearings on the
development of national technological capacity in the LDCs.

National and Local Policies

Government policies have an important influence, both
positive and negative, on the choice and adoption of
technologies. An appropriate policy can raise the supply
of technologies and widen the available range by encouraging their importation and/or indigenous development.
Similarly, it can make some technologies more attractive
to entrepreneurs than others by raising their profitability.
Perhaps even a more important role of the government lies
in providing a socio-political climate and an economic
structure conducive to the widespread use of appropriate
technologies.

Such macro-economic policies as tariff structure, credit
policies, minimum wage legislation, import licensing and
quota systems, influence the direction of research and
development and the pattern of technological innovations.
Many of these policies, though designed and intended for
different purposes, tend to distort factor prices in such
a way that capital-intensive techniques may become more
profitable than the otherwise technically viable and
available employment-generating techniques.

Appropriate policies may include: maintenance of official
exchange rates at their equilibrium value, removal of
interest rate ceilings, tax incentives which favour employment and discourage capital use, etc. In most developing
countries, tax holidays and accelerated depreciation are

introduced with the objective of stimulating investment.
In order to ensure the use of more appropriate employment-
generating technologies, tax rebates may be made condi-
tional on companies providing costing of alternative
labour-intensive production process at the time of replacement or expansion. Choice of a more capital-intensive
alternative in spite of lower costs at prevailing market
prices could disqualify a company for tax rebate.[21]

The impact of government economic policies on the choice
of techniques cannot however be examined exclusively in
terms of factor price distortions. The widespread adoption of appropriate technologies requires an appropriate
industrial structure and inter-firm competition (in mixed
economies) on which governments exercise control through
financial, industrial and trade policies. Thus, although
a favourable price and incentive structure may be a
necessary condition for the use of appropriate technology,
it is not sufficient in itself.

Even where correction of price distortions is desirable,
vested interests whose incomes are likely to be affected
as a result may be so powerful as to block the necessary
policy changes.

Little is at present known about the importance of vested
interests in blocking or facilitating policies that may
not be in their direct interest. Motivations and interests
of different decision-makers - family enterprises, large
private producers, multinationals, public corporations,
etc. - are varied and may often be in conflict. Even when
the economic effects of policies are quite similar, the
distribution of political and social implications for these
groups may differ widely. The prevailing technology-mix
and the structure of production derive from socio-economic
and political objectives of countries. The extent to which
the government policies reflect decentralised production

----

[21]H. Pack, "Policies to encourage the use of appropriate
technology", in USAID, Proposal for a Program on Appropriate
Technology, US Government Printing Office, Washington, July
1976.

structures, "down-scaling", and labour-intensive technical
change, will therefore largely depend on the nature of
these socio-political objectives.

Inquiries into the appropriate links between central and
local planning are also desirable. What types of institutions, governmental and semi-governmental, at the local
level will ensure that the decentralised production is not
hindered by lack of stimulus and resource inputs needed
from higher levels of administration? What types of information are needed by the central planners about the
local resources and needs to ensure that central plans are
consistent with local plans? What are the administrative
and political implications of decentralising planning
from the top? All these questions merit serious consideration.

Appropriate technology development requires policy interventions to stimulate action on both demand and supply
sides. On the supply side, policies concerning R and D
will have to be restructured. In the context of basic
needs strategy, national policy for R and D should give a
much greater priority to small-scale producers, e.g. the
small farmers, the village blacksmiths and other craftsmen,
and rural/small-scale industries in the informal sector.
The objective of the national R and D plan should be to
narrow the technological gaps between the formal (modern)
and informal sectors. This would involve redistribution
of the total volume of R and D resources so that institutions dealing mainly with small-scale and traditional
production receive a much larger share of the total. It
would also imply the location of such institutes in rural
areas closer to productive activity. In other words,
decentralisation of production implied in a basic needs
strategy should go hand-in-hand with decentralisation of
R and D systems and plans. Such decentralisation is
likely to promote greater links between the real needs of
the local community and the R and D activities. Secondly,
it should also encourage the types of R and D which would
tend to be ignored by the central R and D laboratories.

The existing science and technology institutions as well
as the R and D institutes in developing countries, are
concerned mainly with the over-all scientific and technological capacity. They are rarely concerned with labour-
intensive and energy-saving technologies especially designed

for the benefit of the small-scale sector. A few new institutions relating to appropriate technology have also
been established outside the framework of science and
technology institutions. However, establishment of new
institutions in itself will not be enough. It is equally
necessary that the concept of appropriate technology
pervades throughout the governmental decision-making
machinery. The ASEAN proposal to establish national
government organisations with focal points in each ministry
(described earlier) is one of the mechanisms to ensure
scrutiny of sectoral and macro-plans and projects in the
light of alternative technologies.

R and D for the promotion of appropriate technologies at
the local level should also be combined with the creation
of local machine-fabrication facilities, which are
essential for redirecting technical change in the labour-
intensive direction. National policies may have to discourage indiscriminate import of capital goods in order to
promote local production of equipment. Engineering workshops and repair shops in developing countries are often
engaged in shop-floor innovations that are appropriate for
improving traditional technologies. But very little
systematic information exists about such innovations.
Documentation of these innovations in the form of "technical memoranda" relating to specific economic activity to
be written in simple and local languages of developing
countries would go a long way in providing access to
available information on appropriate technologies.[22]

Work on appropriate skills and training for appropriate
technology needs much more attention than it has hitherto
received. National training plans should take account of
potential for on-the-job training and learning-by-doing
effects of decentralised production. Plans and policies
will also be needed at the local levels to introduce tech-

---

[22]The ILO WEP Programme on Technology and Employment is
currently engaged in the preparation of such technical
memoranda in collaboration with developing countries, as
well as with UNIDO and the World Bank. See ILO, Programme
on the Dissemination of Information on Appropriate Technologies, Geneva, February 1978.

nological elements in basic education in secondary schools
and to prepare suitable training materials for schools and
technical centres. Greater efforts will be needed to de-
liver training to scattered and remote rural areas through
mobile training workshops and rural extension services.

The national and local policies have as much to do with
the creation of market for appropriate technologies as for
appropriate products. Policies concerning redistribution
of incomes should stimulate demand for both appropriate
technologies and products. More direct measures may also
be necessary to reduce the importance of "Western-style"
sophisticated products in order to promote simpler goods
with simple characteristics. This may be done either
through restrictions on advertisement and/or restrictions
on the import of such goods. Consumer acceptance of indigenous products and technologies could be promoted
through public information campaigns, appropriate advertisement, and fiscal incentives.

In the context of a basic needs strategy and collective
self-reliance, it may also be necessary for the developing
countries to be stricter about the inflow of private
foreign investment and foreign technology. Domestic innovations, to be successful in their local markets, may need
initial protection against competition from imported technology. Lack of such protection in many developing
countries has partly prevented local innovations on any
significant scale.

Protagonists of basic needs defend such government interventions on the grounds that, in their absence, production
of "basic needs" goods and services is unlikely to occur.
There is no unanimous view on the degree of intervention
or its feasibility. The critics of a basic needs strategy
argue that interventions are often as inefficient as lack
of them. Without entering into ideological issues, we
maintain that a certain minimum of government intervention
is required in a basic needs strategy if only to make sure
that the poor target groups are helped to articulate their
felt needs. It is the responsibility of the governments
to bring appropriate technologies to the poor who may well
be outside the market system.

National policies have an important role to play also in
promoting technical cooperation among developing countries.

Policies to facilitate TCDC could include granting of
special incentives and facilities by governments to their
national enterprises (private or public) which import technologies from other developing countries instead of obtaining them from the advanced countries. The developing
countries often have to pay excessive sums for the import
of technologies from the advanced countries, since the
same technology is often sold and paid for many times. A
TCDC arrangement whereby one developing country government
imports technology from another developing country may in
principle reduce the payment of large sums by the recipient.
Instead of private enterprises acquiring the same technology
several times, the governments could acquire technology and
license its use to private users.

International Policies

International influences on national decision making in
developing countries take place through aid, trade, private
investment and multinationals, etc. The implementation of
a basic needs strategy and reorientation of technological
change in a labour-intensive direction will call for reforms in all these directions. To take the case of the
multinationals first; their main objective is to maximise
profits. This goal is unlikely to be achieved through the
manufacture and sale of "basic needs" goods which can easily
be imitated, are not subject to brand names, and do not
offer scope for rentals that are obtained from proprietary
technology. For the food-processing industry (an industry
of key importance in a basic needs strategy) for example,
brand names and advertising seldom reflect significant
nutritional improvements in products. Indeed, the branded
products are often nutritionally inferior, as is the case
of sifted maize flours. They involve numerous minor
changes in products combined with more elaborate packaging
at increased cost to the consumer. It may be desirable to
discourage such practices of product differentiation by
both foreign and local firms through the taxation of advertising associated with such practices.

There are a number of ways in which multinational companies
could contribute towards R and D in the host countries.
First and foremost, greater effort is required to de-
centralise R and D from the parent companies to their
subsidiaries in developing countries. Host countries could

stipulate that a certain proportion of the net revenue of
subsidiaries of foreign firms be used for local scientific
and technological research. However, this presupposes
that the host country's own R and D priorities favour the
neglected small-scale and traditional sectors. If this
were not so, and if the location of R and D by the multinationals in the LDCs led to an "internal" brain drain
(employment of local scientists and engineers in pursuits
irrelevant to the needs of the LDCs), little could be contributed to the development and implementation of appropriate technologies.

Secondly, the governments of the industrialised countries
should be encouraged to subsidise prices at which multinational enterprises are able to make relevant technological
know-how available to the Third World countries. Such subsidised research should "focus on increasing the ability of
Third World countries to expand employment opportunities,
to satisfy basic needs and to promote self-reliant styles
of development.[23]

Thirdly, efforts need to be made to encourage medium- and
small-scale enterprises in the developed countries to
transfer their technology to the LDCs which could be more
relevant to the latter's needs.[24] These enterprises are
less concerned with world-wide operations. They might
therefore be more willing than the larger companies to
share their technology with the LDCs.

Reorientation of aid policies would also be necessary.
Although, in a self-reliant development path the importance
of aid from the advanced countries may diminish, reliance
on it will not completely disappear. It is therefore
essential that priorities and criteria used by the aid

---

[23]Jan Tinbergen (co-ordinator), Reshaping the International
Order, A Report to the Club of Rome, Dutton, New York, 1976,
p. 155.

[24]A.S. Bhalla, "Small industry, technology transfer and
labour absorption", in OECD Development Centre, Transfer
of Technology for Small Industries, Paris, 1974.

donors should be consistent with the declared goals of
development set by the developing countries. It has been
observed in the past that technological choices embodied
in technical cooperation projects financed by bilateral
aid donors have at times been particularly capital-intensive in situations where labour-intensive alternatives
would have been more suitable. Are such decisions by the
donors irrational, political, inevitable? Do the aid
donors succumb to the prestige factors which are often
attributed to the national governments in developing
countries? Can the information base for technological
choice in project planning and appraisal be improved at
the headquarters of the donor agencies and at the headquarters of the consultants which these agencies use?
Do aid donors finance local costs? Answers to these
questions are currently being sought by the ILO through a
detailed investigation of aid donors' practices, and
through a study of a selected number of aided projects in
the developing countries.

There are a number of ways in which aid can be channelled
more fruitfully towards the application of appropriate
technology. First of all, aid should generally be untied:
it is now well known that local cost financing facilitates
more appropriate technology choices.[25] Secondly, greater
aid should be linked specifically to the development of
national and local technological capability within the
developing countries. This may mean supply of equipment
and staff expertise for science and technology institutions,
support to the on-going appropriate technology centres in
the developing countries in addition to those in the advanced countries which are already receiving adequate
financial support. Thirdly, the bilateral and multilateral
donors supporting the Regional Technology Centres should
link their aid particularly to the promotion of technologies
suited for the production and distribution of "basic needs"
goods.

International trade policies can play an important role

---

[25]In October 1977, the OECD Development Assistance
Committee (DAC) approved a set of "Guidelines on Local
Cost Financing", which is a welcome development.

in ensuring the availability of goods so crucial in a basic
needs strategy. However, we have argued that "Western-style"
products manufactured and consumed in the advanced countries
may not always be relevant for basic needs satisfaction of
the poor target groups. Under these circumstances, trade
among developing countries, particularly in consumer goods,
may be more desirable than trade between LDCs and DCs.

Within the framework of TCDC, a joint action on the part
of developing countries will be necessary to establish
regional marketing organisations, improve transport facilities and product quality in order to enable export of
products, especially those manufactured by small-scale
industries, to neighbouring countries.

The developing countries could also promote the adaptation
and use of advanced technologies by creating multinationals
of their own in such fields as energy, transport and pharma-
ceuticals.

Another area for international action is the "unpackaging"
of technology imports. Ability to import different elements
of technology from different sources would enable the
developing countries to adapt technology to local uses and
to benefit from the learning process involved in such
adaption. International agreement in the form of a code
of conduct on technology transfer (work on which is in
progress under the auspices of UNCTAD) should help to bring
about such unpackaging.

CONCLUDING REMARKS

We have stated in this paper that a basic needs strategy
is a more comprehensive view of development. It is a
strategy whose objectives have been somewhat redefined to
concentrate on the material and non-material welfare of
those target groups who are below the average for the
economy as a whole. One can just as well argue that the
objectives of earlier development strategies have been more
or less similar, if not the same: it is the instruments to
achieve these objectives and the time horizon that are
altered in a basic needs strategy. For example, as much
emphasis is placed on consumption planning as on production
planning, more stress on decentralised production than on
central control, on popular participation than participation

only by the privileged few, on redistribution of incomes
and assets rather than purely on fiscal incentives. The
fulfilment of basic needs of the bulk of the population
is likely to be achieved only marginally through price
and market mechanisms alone. Reforms in the incentives
structure must be accompanied by the necessary institutional reforms.

Appropriate products and appropriate techniques will both
require a pride of place in an appropriate technology
strategy for a needs-based development. Technological
self-reliance essential in such a strategy would require
greater national political commitment and greater technical
co-operation among developing countries. International
action is designed only to assist the developing countries
to implement their goals of development for their societies.
Unless national and international policies and action programmes are developed in harmony, there is not much hope
of implementing the declared goals in the foreseeable
future. This implies that the objectives of basic needs
policies (largely a national concern of the LDCs) should
be consistent with the establishment of a New International
Economic Order (a joint concern of the LDCs and DCs).

PART II
Existing Institutional Framework

Chapter 3
NATIONAL AND REGIONAL TECHNOLOGY GROUPS AND INSTITUTIONS: AN ASSESSMENT

A.K.N. Reddy

INTRODUCTION

Appropriate technologies can be defined as those technologies which advance the socio-economic objective of
development, the latter being viewed as a process which
is primarily directed towards:

(a) the satisfaction of basic human needs (starting
    with the needs of the neediest, viz., the urban
    and rural poor);

(b) endogenous self-reliance through social parti-
    cipation and control;

(c) harmony with the environment to ensure the
    long-term sustainability of this development
    process.

---

[1]Constructive comments on the first draft of this paper
were received from Frances Stewart and Ajit Bhalla, to both
of whom the author wishes to express his sincere thanks.
The author also wishes to thank M.K. Garg and J.N. Powell,
and Ajit Bhalla, who subjected the second draft to a number
of useful criticisms.

[2]Professor, Indian Institute of Science, Bangalore, (India);
Convener, ASTRA (Cell for the Application of Science and
Technology to Rural Areas), Indian Institute of Science; and
Secretary, Karnataka State Council for Science and Techno-
logy, Bangalore.

All the three crucial aspects of appropriate technology,
viz., choice, generation and dissemination, must be inter-
linked, possibly through a mechanism such as shown in
Fig. 1. Some aspects of this inter-relationship are
discussed below.

In principle, the process of technology diffusion within
one country can be guided by general guidelines which can
be derived from the experience of other countries; but, in
practice, effective guidelines - and the corresponding
institutions to implement these guidelines - are best generated endogenously to suit the traditions, institutions,
skills and history of the particular local, sub-national
and national environment. The next best alternative is to
utilise guidelines from similar countries in the sub-region
and region, and from institutions covering these areas.

Implemented technologies have inevitable impacts on the
process of development in general, and on the lives of the
urban and rural poor in particular. Thus, a monitoring of
these impacts must influence the criteria for the choice of
appropriate technologies - the criteria must be modified
and improved, or validated and confirmed. These improved
and/or confirmed criteria must be linked, on the one hand,
to the choice of technology, and on the other, to the development of technology.

The process of choosing appropriate technologies involves
the screening of a bank or list of available technologies
with the aid of improved and/or confirmed criteria.

In so far as the criteria must be developed in close
association with the dissemination of technologies, and
with the monitoring of the impacts of this dissemination,
it follows that the effectiveness of the technology selection process is enhanced to the extent that it is done at
the national level, and perhaps even at the sub-national
and local levels.

Neighbouring countries in the sub-region and region, as
well as distant countries facing similar developmental
tasks, and institutions dealing with these countries can
play an important role by supplying information on available
technologies, and thereby enlarging the national bank of
technologies from which appropriate technologies are chosen.
Nevertheless, the main thrust for the enlargement of the

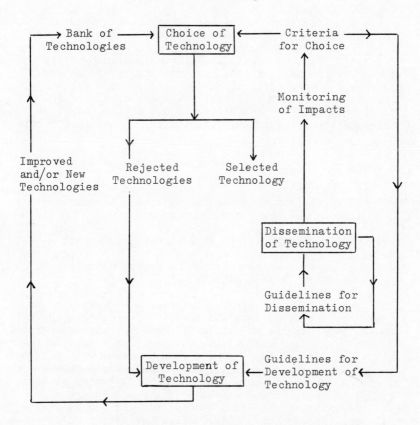

Fig. 1: <u>Inter-relationship between choice,
development and dissemination of
appropriate technology</u>.

bank of technologies must come from the internal development of technologies, particularly because this process of
technology development must be coupled closely, on the one
hand, with the choice of technologies, and on the other
hand, with the dissemination of technologies.

The above discussion leads to the conclusion that overwhelming emphasis must be placed upon the development of
national capability in the selection, generation and
diffusion of appropriate technologies, with external inputs to these three processes playing a supportive or
catalytic role. Not only is this national capability
essential for the ultimate effectiveness of these processes of choice, generation and dissemination; it is the
inescapable basis for the self-reliance of countries. In
fact, this national capability should be the essential
precondition for international action to play an effective
role.

In this context, three crucial problems must be posed:

(a) Does national capability for the generation
    and dissemination of appropriate technologies
    exist in the developing countries?

(b) Upon what factors does this capability depend?

(c) What steps need to be taken to generate and/or
    strengthen this capability?

The purpose of this paper is to present a discussion of
these problems. In order to initiate the discussion with
an empirical basis, an outline of some on-going activities
in appropriate technology is provided. This outline includes an indication of some of the problems with directories of appropriate technology organisations; it also
contains a brief description of a limited number of these
organisations. An assessment of their efforts requires
the use of valid criteria, and in order to generate such
criteria, models for the development of appropriate technology and for its dissemination are suggested. A set of
tentative criteria are then derived from these models.
These criteria are used to arrive at a preliminary assessment of institutions dealing with appropriate technology.
Some critical shortcomings and limitations of national
efforts in the development and dissemination of appropriate

technology are also highlighted. The identification of
these bottlenecks leads to the recommendations for inter-
national/global action.

APPROPRIATE TECHNOLOGY GROUPS AND INSTITUTIONS

Appropriate technology is engaging the attention of a very
large number of organisations. Several lists of such
organisations have been prepared, and still others are
under preparation. For example, lists have been prepared
by TRANET (Transnational Network for Appropriate Technologies, USA), by the Commonwealth Secretariat, and by
ITDG (Intermediate Technology Development Group, Ltd.,
London) on behalf of the United Nations Environment Programme. Also, the ILO publication on "Technologies for
Basic Needs" (1977) includes an appendix on "Institutions
dealing with Appropriate Technology".

To indicate the dimensions of such lists, the first version
of the ITDG-UNEP directory refers to about 50 appropriate
technology organisations in Asia (including the Middle-East
and the Far-East) and about 75 in Africa, with a worldwide
"head-count" of about 275 organisations.

There are several problems with such lists. Firstly, they
are overwhelmingly based on explicit declarations of
interest in appropriate technology. But, like the character
in Moliere's play, some institutions may be working on
appropriate technologies without being aware of it, in which
case they may not find a place in the list. Also, many institutions which are included in the list may turn up with
technologies appropriate for rural areas rather than technologies specifically appropriate for the rural poor, when
in fact the latter are only a sub-set of the former. Thus,
the question of which institutions to deem as "appropriate
technology institutions" needs to be resolved.

Secondly, the question of lists is complicated by the fact
that there is a growing appropriate technology "movement"
as part of the "counter-culture" in developed countries;
and many groups which cannot easily find a place in the set
up of conventional technology, enlist in this new movement.
Further, the presence of such "off-beat" groups often repels
conventional institutions which may otherwise have far
greater potential for generating and disseminating appro-

priate technology. Conversely, the appropriate technology
"movement" often tends to exclude established institutions
of education, science and technology on the grounds (invariably justified!) that such institutions are predominant-
ly concerned with western technology. The compilation of a
directory of appropriate technology institutions is, therefore, not such a simple and objective matter.

Thirdly, the question of the potential for generating and
diffusing appropriate technology is a crucial one. It may
be as important to encourage (with suitable measures)
established institutions of education, science and technology to commit themselves to appropriate technology as to
buttress groups which can only achieve limited success.
This is particularly important in those developing countries
which have already built up significant systems of education,
science and technology.

Consider the case of India for example. The country has 115
institutions of university status, 44 national laboratories
of the Council for Scientific and Industrial Research, 28
laboratories of the Indian Council for Agricultural Research
and 8 laboratories of the Indian Council for Medical Research with a total of over 10,000 personnel with post-
graduate qualifications. Despite this, the usual appropriate technology list only mentions 17 organisations in
India with perhaps about 250 qualified personnel altogether.
Thus, assuming that "AT potential" can be gauged by the
number of trained personnel, the Indian potential for
generating appropriate technology must be at least 50, and
perhaps 100, times that suggested by the above-mentioned
directories of appropriate technology organisations.

This type of discrepancy is a crucial issue to international
organisations which are striving to promote national capability for the generation and dissemination of appropriate
technology - should the "AT potential" of a country be
judged by the mainstream of education, science and technology, or by the list of appropriate technology organisations however far from the mainstream these organisations
may be? Of course, an either-or approach may be quite misleading, and the actual and potential contributions of both
groups of organisations must be harnessed.

Without pre-empting the resolution of this important issue,
one important conclusion can be drawn: the conventional

institutions of education, science and technology in
developing countries are contributing far less to appropriate technology than is commensurate with their potential.

Focussing on the appropriate technology institutions
mentioned in the lists, it is quite obvious that a description of all these institutions would be quite pointless.
Hence the description of a limited sample of ten institutions has been assembled below in order to initiate discussion on the assessment of appropriate technology activities. It is neither claimed nor intended that the sample
covers all geographical regions and/or the most effective
appropriate technology organisations.

1. Appropriate Agricultural Technology Cell (AATC),
    Bangladesh.[3]

"This Cell was established in 1975 under the administration
of the Bangladesh Agricultural Research Council. The objectives of the Cell cover: development and promotion of
labour-intensive and capital-saving machinery and tools and
implements for agricultural production, manufacture of implements through greater utilisation of local resources,
development of appropriate drying, storage, processing and
milling facilities to prevent post-harvest losses".

"The principal activities that the Cell undertakes are the
collection and dissemination of information on appropriate
technologies for the rural sector and initiation and promotion of research on rural technologies through grants to
researchers in universities and other institutions. Working groups are established in the following fields: draught
power, irrigation, fertiliser uses and agronomical methods,
post-harvest operations, and agricultural workshops".

"Although this Cell is at present small with limited number
of projects, it is proposed to expand work into such fields
as animal husbandry, village-based industries, rural
housing, etc. A proposal for an autonomous institute of

---

[3]See Hans Singer, Technologies for Basic Needs, Appendix
B: "Institutions dealing with appropriate technology" ILO,
Geneva 1977.

Appropriate Agricultural Technologies is also under consideration."

2. <u>ASTRA (Cell for Application of Science and Technology to Rural Areas), Indian Institute of Science, India.</u>

ASTRA was created in 1974 within the Indian Institute of Science (which is one of the oldest and prestigious institutions in the country) in order to serve as an agency for increasing the Institute's awareness of rural problems; and play a key role in correcting the present urban bias in the educational, research and development programmes of the Institute, so that a significant fraction of these programmes acquire a rural orientation. ASTRA's programme is concerned with the development and promotion of appropriate technology for the satisfaction of basic needs, defined in terms of access to inexpensive essential goods and services for the unemployed and underemployed rural poor.

The first phase of the programme of activity includes:

"(a) the development and testing of village-oriented technologies on the Institute campus;

 (b) the establishment of an Extension Centre in a village near Bangalore; and

 (c) the transfer of developed and tested technologies either to the village through the Extention Centre or to other rural development agencies."

The approach used is not only to derive appropriate technologies for the rural poor by merely simplifying the modern techniques used in urban areas, but also to start from the observation and study of rural traditional techniques, and therefrom to improve them and increase their efficiency. The initial phase of ASTRA thus involves extensive grass-roots learning and field surveys in order to identify the most crucial problems of the rural poor and the technical solutions to be investigated.

In three years of existence, ASTRA has grown as an active inter-disciplinary group working on a wide range of problems relevant to rural areas.

The following four categories of ASTRA's work may be
mentioned:

(a) Sponsored work, in which 9 projects (biogas technology,
    windmills, bullock carts, rural energy consumption
    patterns, community biogas plants, bamboo conservation,
    energy planning, hand pumps and village ecosystems)
    involving 17 faculty members and 15 project assistants
    have been supported to the extent of about $100,000 for
    periods ranging from 9 to 36 months by 4 agencies (Tata
    Energy Research Institute, Department of Science and
    Technology, Indian Council for Social Science Research,
    Karnataka State Council for Science and Technology);

(b) Faculty research, in which 7 topics, including vapour
    pulse pumps, silk worms, alternative energy sources,
    small-scale soap production, low-cost building con-
    struction, rural housing and educational aids for
    science teaching, are being investigated by 8 faculty
    members;

(c) Student dissertation projects, (as part of the Master
    of Engineering course requirements) in which there are
    11 investigations on modelling of biogas units, sodium
    silicate from rice husk, plastics from castor oil,
    edible cellulose from rice husk, cellulose fibre from
    groundnut shells, energy survey of building materials,
    rammed earth construction, soil cement blocks and
    stabilisation, solar airconditioning and heat pipes;

(d) Ungra Extension Centre work, in which the design of all
    the new ultra-low-cost buildings (dormitory, library-
    cum-office, faculty and labour housing, seminar hall)
    has been completed and construction work has started,
    the survey of rural energy consumption patterns is
    almost complete; a windmill has been installed, and the
    study of a village as an ecosystem has commenced.

The highlights of ASTRA's work during 1975-76 are:

(a) The successful identification, in collaboration with
    the Karnataka State Council for Science and Technology,
    of the causes of widespread failure of handpumps for
    village drinking water borewells, the suggestion of
    modifications to prevent these failures, and the
    successful field testing and dissemination of these

modifications;

(b) The successful development and field installation of an
innovative, low-cost, vertical-axis, Savonius type
windmill for water pumping; and

(c) The construction on the Institute campus of a low-cost
building, and its regular use as a laboratory for the
study of solar airconditioning and biogas technology.

3. Appropriate Technology Development Organisation (ATDO),
Pakistan.

ATDO came into existence in July 1974. It was originally
attached to the Ministry of Science and Technology which
transferred it to the Planning Commission in April 1975.
Some nationalised banks came forward with funds and help to
enable ATDO to start its development work. To popularise
the possibilities of development based on appropriate technology, ATDO organised in March 1975 an exhibition in which
a low-cost house, hand-made match manufacture, a biogas
plant, a high-extraction screw-type cane-crushing machine,
and other appropriate technologies were displayed. Development work was also initiated on the manufacture of hand-
made paper and of paper pulp from banana trunks, and on
biogas plants and windmills.

During its second year of operation, ATDO was able to secure
office accomodation, but technical staff became available
only towards the very end of the second year. Continued
funding from nationalised banks permitted further progress
on the projects which it had taken up during the first years,
as well as the initiation of work on new projects, such as
under-soil irrigation through earthen pitchers and PVC pipes,
low-cost housing and primary schools, ox-driven implements,
paddy dryers, insecticides from paddy waste and simple low-
cost hydroelectric plants.

The third year of ATDO's existence is significant for four
reasons:

(a) In April 1977, ATDO was declared an autonomous body in
order to overcome a number of procedural, financial and
administrative problems which were making its task of
technology development extremely difficult;

(b) The dissemination of a few technologies, for example, hand-made match manufacture, hydroelectric generating plants based on water wheels, and screw-type cane-crushing machines, was initiated;

(c) Development work on some other technologies, for example, roofing for low-cost housing, windmills and under-soil irrigation, reached the stage of completion;

(d) ATDO also commenced a number of new R and D programmes on ferro-cement boat-building, rural assembly of trans-istor radios, candlesticks and chalk-stick manufacturing, hand-operated multi-spindle spinning machines, hand-made paper manufacture, etc.

A striking feature of ATDO's approach is the great stress laid on the importance of people's participation. By spreading the concept of appropriate technology and de-monstrating such technologies, it is believed that the people can be motivated to undertake development for them-selves. ATDO has even advertised in the daily newspapers calling for suggestions from the people - and incidentally secured an overwhelming response to this People's Participa-tion Scheme. Another example of ATDOs success with popular participation in the development of appropriate technology is the technical contributions elicited from a local black-smith in the designing and fabrication of an animal-drawn Fresno Scraper for land-levelling and earthmoving.

With the achievement of autonomy, the stage seems set for even more significant contributions from ATDO, provided that it can stimulate and sponsor high quality technical work on a much larger scale.

4. Council of Scientific and Industrial Research (CSIR), India.

The CSIR which was founded in 1944 is today one of the major scientific agencies in India with an annual expenditure of around $30 million in 1974-75. Under its purview are 44 re-search laboratories manned by about 5000 scientists working in areas ranging from aeronautics and electronics to food technology and environmental engineering. Whereas its basic orientation has been towards indigenous industry which it has assisted in the drives for import-substitution and greater productivity, the CSIR embarked on a major rural development project in 1974. Project Karimnagar is an experiment where the development of a backward area (viz., the district of Karimnagar of area 11,800 kms and population 1.96 million) is being attempted through the application of science and technology. The aim is maximum utilisation of available resources following an integrated approach. Marshalling expertise and technologies from its various constituent laboratories, CSIR is implementing an integrated development plan for appropriate technologies in agriculture (in collaboration with the Indian Council for Agricultural Research), public utilities such as roads, housing, buildings and public health and industry (agro- and mineral-based industry). The Karimnagar experiment is now undergoing a rigorous evaluation for mid-course correction, if necessary.

5. Development Technology Centre (DTC), Institute of Technology, Indonesia.[4]

"The Development Technology Centre (DTC) is a flexible, self-supporting organisation based at the Institute of Technology Bandung (ITB) in Indonesia. Established in 1973 by a decree of the Rector of ITB, DTC consists primarily of ITB professors and staff members who are interested in conducting research and development programmes related to national development."

"The focus of DTC activities is the application of a wide range of appropriate technologies to meet the challenges of unemployment and underemployment in Indonesia. Priority concerns are the planning, selection and development of appropriate technologies and the specific skills necessary for integrated development".

"DTC programmes have received support from various sources, including Indonesia Government offices and banks, international agencies and foundations, local governments and organisations, and institutions of higher learning".

---

[4]See SIDN Newsletter, Vol. 3, No. 1 (1976) published by the Georgia Institute of Technology, USA.

"Currently, DTC is involved in an extensive five-year programme in rural appropriate development technology in collaboration with the TOOL Foundation of the Netherlands.
The programme will establish a technical information system;
hardware development projects; a system of field stations
for realistic testing and demonstration of hardware or
software technologies; and an inquiry and extension activity
complemented by publication, documentation, and training
programmes. Also in progress is a joint study with the
Council for Asian Manpower Studies on the relationship
between local and small industries and a multi-national
joint venture emphasising vehicle assembly."

"An important aspect of DTC's work is the training and
development of entrepreneurs through achievement motivational
training programmes."

"DTC also works on development of local power sources -
solar energy, wind energy, bio-gas, micro-hydro, and
integrated systems - mainly for the rural areas and in the
form of autonomous or decentralised systems."

"Some sample projects in appropriate technology hardware
development and testing are a nonelectric ice maker, an
agricultural product dryer, a stone cutter for the cottage
jewelry industry, and food processing technologies such as
a Kemiri nut shelling machinery and coconut processing".

"Another area of activity involves technical needs assess-
ment, regional development, and technology transfer studies.
The goal is regionalisation of technology transfer through
Regional Development Technology Centres."

"DTC presently operates a field station for the purpose of
appropriate technology development and demonstration in
and around Jogjakarta, Central Java."

6. East African Industrial Research Organisation (EAIRO),
   Kenya.[5]

The organisation covers three East African countries, namely

---

[5]With the break-up of the East African Community, this institution has been renamed as the Kenyan Industrial Research
Organisation.

Kenya, Uganda and Tanzania, and is mainly oriented towards small-scale production units in the primary and secondary sectors. In particular food programmes in agriculture receive special attention through R and D efforts to promote appropriate cultivation in different geographical areas (e.g. substitution of sorghum and millet for maize in semi-arid regions). Technical innovations are also carried out in the industry (brickmaking, food-processing, energy).

The work of EAIRO is not biased in favour of the large-scale organised sectors as is often the case of many national and regional research institutions. Instead, it is in line with the basic needs approach. Many of the smaller research projects are initiated in response to demands from small-scale clients.

Some of the specific examples of innovations are:

(a) Development of a solar water heating system which can be manufactured locally by small-scale sheet metal enterprises and which could provide hot water for domestic use in rural areas;

(b) Reconditioning of disused kilns in the ceramics section for the manufacture of bricks and tiles;

(c) Development of techniques for the commercial manufacture of oriatiô, a natural dyestuff used in some dairy products;

(d) Development of techniques for using Kiisi stoneware for electrical insulators."

7. ESCAP Regional Centre for Technology Transfer (RCTT), India.[6]

The decision to establish a RCTT in India was the result of almost a decade of consideration, consultation and re-

---

[6]See "Project Document on Regional Centre for Technology Transfer" (E/ESCAP/44/Add. 1, 7 March 1977).

commendation by the international community. Of the nine
immediate objectives envisaged for such a RCTT, the following make specific reference to the development and dissemination of appropriate technology:

(a) "To set up a suitable clearing-house for intra-regional
    and inter-regional exchange of information and ex-
    perience relating to technology development, adaptation
    and transfer, and to promote cooperation in such
    activities, including the joint adaptation and develop-
    ment of appropriate technologies";

(b) "To assist in carrying out studies on selected tech-
    nological problems and on development of appropriate
    technologies of interest to several countries of the
    region."

The case for a RCTT rests on the crucial importance of
developing countries having national centres that are concerned with technology development, adaptation and transfer,
and _inter alia_, with promoting "interest in the concept of
"appropriate" technology among government policy-makers and
adminstrators, industrialists in public and private sectors,
entrepreneurs large and small and the staff of the universities and technical institutes". Whereas some developing
countries already have institutions and organisations which
may be considered incipient versions of national centres, a
large number have not even commenced the building of the
required infrastructure. Thus, an immediate objective for
the RCTT is "to promote the establishment of national
centres" and "help strengthen their capabilities in this
regard". This means that a crucial objective of the RCTT
is "to function as the lynchpin of a network of national
centres to be set up in individual countries of the region".
In the specific matter of appropriate technologies, it is
envisaged that the RCTT will "promote the exchange of information and experience on such technologies and their
transfer among countries in the region", and also "promote
and organise regional cooperation in research and development of technologies appropriate to several countries of
the region".

In short, "the functions of the RCTT and of the national
centres can be analysed into two broad categories: technology information; and technology evaluation, adaptation
and development. A third and equally important function

is the concept of using the regional centre for ...
sponsoring research into basic technologies required
by the region."

Though there are specific and explicit references to
appropriate technologies in the objectives and functions
envisaged for the RCTT, it is clear that appropriate
technology may not necessarily be its sole, or even
predominant, concern. Further, there are possibilities
of conflict between elements of the mandate - for example,
the objective of promoting "the transfer of technologies
adopted by developing countries within the region" can be
inconsistent with the objectives of promoting appropriate
technologies, if the technologies which have already been
adopted are inappropriate.

Thus, the RCTT has potential for the development and
dissemination of appropriate technologies, but the
realisation of this potential depends largely on the
emphasis placed on the different objectives and functions.
At this stage, it is too premature to judge the issue
because the RCTT is just in the process of being
established.

Thus far, its organisational structure has been delineated -
it will have a 14 member Governing Board, a Director who
will be advised by a Technical Committee composed of the
directors of national centres, and three divisions, viz.,
Technology Information, Technology Evaluation and Development, Management Personnel and Training, with supporting
staff of professionals. A 34-month work programme -
including pre-operational and operational phases, has been
drawn up with a starting date in February 1977. ESCAP has
been selected as the executing agency. Both the site for
a permanent home, as well as a building to serve as a
temporary home, for RCTT have been located in Bangalore,
India, and the project is operational.

8. International Rice Research Institute (IRRI),
    Philippines.[7]

The International Rice Research Institute (IRRI) was

---

[7]See CGIAR - Consultative Group on International Agricultural Research, (undated).

established in 1960 to conduct research in all aspects of
rice production, and in particular the development of improved rice strains and rice farming methods. "IRRI won
early celebrity with the development of IR8 and the host
of semi-dwarf rices that soon followed." These "varieties
rapidly became the most widely grown in the tropics, and
today about a fourth of the world's rice land is planted
to semi-dwarf rices of the IR8 type."

"The technological advances that doubled yields in some
places were developed by inter-disciplinary teams concentrating primarily on genetic manipulation of the
tropical rice plant. Agronomists, pathologists, entomologists, geneticists and other scientists worked together
to produce a range of high-yielding rice varieties to feed
more people from the same land". IRRI scientists achieved
this remodelling of the rice plant (a) "by collecting and
screening thousands of varieties of rice from across the
world", and (b) manipulating, through cross-breeding, the
genes that control each favourable trait.

When the first IRRI rices were proved in experimental
plots, "the Institute set about developing extension and
demonstration techniques to get these rices into the hands
of farmers and to teach farmers how to grow and protect
them." What has come to be known as a "package of
practices" was developed, in which all the inputs (seed,
fertiliser, insecticide) and the instructions for their
proper use, were elaborated. There was also the invisible
part of the package, viz., the institutional support involving technical assistance from extension agents, credit
through government programmes, guaranteed selling price,
etc.

In its first ten years, IRRI has had a substantial impact
on rice production in the developing countries. It
realised, however, that many major problems remained to be
solved. The most outstanding problem concerns the fact
that "despite significant increases in rice productivity
in areas where farmers are assured of water control and
chemical inputs, the new rice technology has bypassed
most less prosperous areas." In many of these areas, the
semi-dwarf rices are too short to grow in the vast deep-
water regions along the mighty rivers. Similarly, high-
yielding rices are needed for the salty soils of coastal
marshes and of irrigated land in arid regions, and for the

drought-prone regions where upland rice is grown. Also,
"the improved rice developed for all areas must be resistant
to major insects and diseases."

To meet this challenge, IRRI has formalised an institute-
wide Genetic Evaluation and Utilisation (GEU) programme as
"an inter-disciplinary rice improvement effort, linked with
national programmes in Asia, Africa and Latin America, to
jointly develop and evaluate improved rice and technology
for all rice-growing areas. Nine inter-disciplinary teams
of plant breeders and problem-area scientists, such as
pathologists, entomologists, physiologists, and soil and
cereal chemists, work together to develop rices that are
genetically adapted to "agronomic characteristics; re-
sistance to insects, diseases and drought; tolerance to
adverse soils, deep water, floods and extreme temperatures;
grain quality; and higher levels of protein. "To develop
improved rices, each team first identifies varieties that
have other favourable traits. The progeny of these crosses
are tested under severe stresses, so that scientists can
select experimental lines that can withstand harsh conditions." This work is actively in progress, and the results
achieved thus far are promising.

IRRI also "collaborates with economists and agronomists in
rice-growing countries to develop a methodology for the
monitoring of problems that slow down the farm adoption of
improved rice varieties and technology. Scientists conduct
experiments on farmers' fields, survey farmers to determine
biological and socio-economic constraints and analyse
markets and input prices." "Agro-economic teams seek
answers to such problems as why rice production has substantially increased in many new regions where the new
varieties are planted, but not in others. Or why many
farmers who have accepted the new rice varieties still do
not use accompanying chemical inputs. Once answers are
determined, scientists can tailor research to develop
varieties and technology to overcome the production constraints."

To intensify food production, farmers in developing
countries need tools and technology to speed up certain
agricultural operations, such as land preparation,
threshing and drying. But "many of the machines designed
for large-scale farming in the developed countries are too
costly and complex for farmers in the rice producing

countries. Besides, they are not easy to service and
maintain because spare parts are scarce and expensive.
Finally, the machines cannot be economically manufactured
in low volume in developing countries because they are
designed for capital-intensive mass production."

IRRI has sought to tackle these problems through its Farm
Machinery Development Programme which was started in 1965.
The programme aims at developing farm machines that satisfy
two major conditions:

(a) Designs must be compatible with the technical
    and economic needs of small farmers who use them;

(b) The manufacture and servicing of the machines must
    be within the technical capabilities of indigenous
    small and medium-scale machine shops. IRRI gives
    drawings, designs, and limited technical support
    free of charge to manufacturers. By 1975, about
    11,000 IRRI-designed machines, including its 5-7HP
    power tiller, axial flow thresher, batch dryer, and power
    weeder were commercially produced by small manu-
    facturers in Asia. These machines are meant primarily
    for the 2-10 hectare farms, it being assumed that the
    traditional manual and animal-drawn farm implements
    are adequately serving the needs of the less-than-
    2-hectare size farms. If this assumption is not
    valid, the IRRI Farm Machinery Development Programme
    will bypass the poorest farmers who can benefit most
    from improvements in productivity.

To enhance the programme's effectiveness, IRRI established
in 1976 regional industrial extension offices in Pakistan
and Thailand. The IRRI-PAK Agricultural Machinery Programme, for instance, aims at introducing IRRI-designed
machines to farmers and manufacturers in Pakistan and the
neighbouring countries. Since, however, most of the IRRI
machines were originally developed for wet-land farming
practices, the IRRI-PAK programme is focussing on modifying
and adapting these machines for dry-land farming conditions.
Currently, the emphasis is on disseminating the axial flow
thresher, the root-zone liquid applicator and the diaphragm
pump.

IRRI's experience has helped chart a new course in institutional development. It influenced the subsequent

formation of the other agricultural institutes under the
programme of the Consultative Group on International
Agricultural Research (CGIAR) as well as their nature,
staffing patterns and directions of programmes.

There is, however, another perspective from which the
performance and effectiveness of IRRI should be examined.
IRRI is located on grounds adjoining the campus of the
University of Philippines, Los Baños, (UPLB) which has
done pioneering work in rice research. Several questions
arise: (a) could not the achievements of IRRI been attained
by UPLB if the latter had been given selective and critical
support so that its activities acquired an international
dimension ? ; (b) would not this alternative strategy of
introducing a major international component into UPLB's
rice research have involved far less investment?;
(c) what effect has the presence of an international institution like IRRI with its lavish equipment, international
salaries, etc., had on the morale of UPLB - has IRRI
stimulated UPLB to greater heights or overawed it into a
psychology of inferiority and ineffectiveness? Detailed
studies on these questions have not yet been carried out,
but first impressions indicate that international institutions like IRRI undermine the self-confidence of
national institutions and have a debilitating effect on
national capability.

9. Korea Institute of Science and Technology (KIST),
    Korea.[8]

"The Korea Institute of Science and Technology (KIST) is
a large, multi-disciplinary contract research organisation
located in a metropolitan city of Seoul, Korea. It is a
wholly autonomous and self-sustaining institution engaged
in the research and development of science and technology
for the benefit of the Korean ... industry and economy ...
KIST was formally created in February 1966 and the dedica-

---

[8]See Nam Kee Lee, Technological Development and Role of
R and D Institutes in Developing Countries - The Korean
Case, World Employment Programme Research Working Paper
(WEP 2-22/WP.25), December, 1975.

tion ceremony ... took place in October 1969 ... Six years after KIST initiated its R and D activities", it was viewed as "a viable research institute".

"KIST is unique in several aspects; its threshold size is adequate enough to include diversified areas of technology and to conduct effective research works; its research laboratories are rightly sized, well-equipped, and adequately funded and so is competent enough to compete with foreign source of technology. Government support has been consistent and adequate enough to overcome major difficulties ... The input expenditures for KIST to date amount to a total of $24.1 million to build, equip and endow the Institute."

"KIST, as the largest R and D institute in Korea, has played an important role in the transfer, adaptation and development of appropriate technology in Korea's bid for rapid industrial development ... KIST has adopted the contract research system by which KIST is held responsible for the execution of the required research under contract with local sponsors, and for the submission of progress reports periodically. After the study is over, all research results as well as supporting data, information and patents become the sole property of the sponsor."

"The research activities at KIST are organised to cover six general subject areas such as mechanical engineering and metallurgy, electrical and electronic engineering, chemistry and chemical engineering, food and feed research, techno-economics and other supporting services including computer service department, technical information services, chemical analysis laboratory, material testing laboratory, machine shop, pilot plants and library. In 1974 KIST performed contract research work on a total of 204 items: electrical and electronics accounted for 17 items (representing 16.9 per cent of total research contract amount in 1974), mechanical engineering 17 items (17.3 per cent), chemistry and chemical engineering 36 items (12.8 per cent), food and feed research 6 items (6.2 per cent), metallurgy and materials 18 items (12.2 per cent), and miscellaneous services 96 items (34.6 per cent) respectively."

"The staff of KIST includes 852 members as of September 1975 in which four (0.5 per cent) are in top management, and 196 (23 per cent) in research ... In addition to above regular staff, about 100 part-time investigators, engineers and advisors are employed to help the KIST research activities."

"As a result of KIST's efforts to develop new products and new processes for the local clients a total of 131 patent applications had been filed by the end of 1973 in which 13 were the applications for foreign patents."

"What has been accomplished so far at KIST has clearly indicated that the major role and activities of the Institute have been heavily concentrated on (a) the identification and selection of appropriate technology for the local clients, through its technical survey and feasibility studies; (b) the adaptation of existing technologies to meet specific local needs through applied research and technical services; (c) the development of appropriate and relevant technologies to suit the intrinsic conditions of local industry through indigenous research activities."

In 1978, KIST is altering its main thrust of activity. Having jargely "delivered the goods" as far as Korean industry is concerned, it is defining new missions. In particular, it is involving itself in Korean science and technology planning, and in the Saemaul Undoug (New Village Movement) programmes of the government. With regard to the latter, it is engaged in a major experiment in Cheju island. Assembling a large inter-disciplinary team, it is addressing a wide variety of rural settlement problems, including alternative sources of energy, housing, sanitation, water management, etc. This commitment to a rural region and to Saemaul Undoug is bound to transform the character of KIST and make it a leading organisation for rural appropriate technology.

10. Technology Consultancy Centre (TCC), University of Science and Technology, (Kumasi), Ghana.[9]

"The Centre was established in 1972 to serve as an inter-

---

[9]ILO, Technologies for Basic Needs, op. cit., Appendix B.

mediate between the University specialists and the
potential users in the public, and has become largely
involved in small-scale industries. The Centre participates
in the research efforts by providing technical know-how and
assists in the testing of new products in pilot plants. It
also provides technical assistance to firms in terms of
quality control, commercial production, access to credit
and equipment improvements."

"The Centre has developed a reputation for stimulating
grass-roots development through the application of intermediate technology. Some of the examples of such work
include the upgrading of existing craft industries such as
textiles, wood-working and pottery. The development of
appropriate processes includes: manufacture of spider glue
from cassava starch and alkali from plantation peel - the
raw materials which are in abundant supply in Ghana; and
manufacture of broad-looms for village weavers. In the
case of the manufacture of glue, the Centre provided
technical know-how, production plant and a financial loan
to the entrepreneurs. In addition, the Centre has established three production units on the University campus for
the manufacture of nuts and bolts, soap bars and broadlooms
for weaving. A soap pilot plant is the largest single
project of the Centre which is engaged in commissioning
seven small-scale soap-making plants (200-500 bars per day)
using mostly local raw materials and serving rural markets."

"Recently, a programme has also been launched for the
establishment of craft centres in some 40 Ashanti villages.
Other rural non-farm activities include glass bead-making
coconut products, brass casting, and the local manufacture
of such agricultural equipment as pumps, driers, and
bullock carts."

*   *   *

A limited number of institutions dealing with appropriate
technology have been briefly described above. The task now
is to assess whether such institutions have the capability
to develop and disseminate appropriate technology. Such
an assessment, however, should not be done arbitrarily; it
must be based on a valid set of criteria. Since, however,
there is no accepted set of criteria, an attempt will be
made below to present a model for the development and
dissemination of appropriate technology, and to derive
therefrom a tentative set of criteria with which to assess
appropriate technology institutions.

TOWARDS A FRAMEWORK OF ANALYSIS

Development of Appropriate Technologies

A fundamental assumption underlying the framework to be
proposed here is that the pattern of technology is shaped
by, and in turn shapes, the society in which this technology is generated and sustained. More specifically,
technology responds to social wants[10] which are in turn
modified and transformed by technology through a causal
chain, or rather causal spiral, which can be schematically
represented as shown in Fig. 2.

Some features of the conceptual scheme represented by
Fig. 2 are elaborated below.

(1) Though the majority of the innovations underlying the
industrial revolution of the eighteenth and nineteenth
centuries came from craftsmen and artisans working outside
the framework of formal institutions of learning, the
present situation is quite different. Today, it is the
institutions of education, science and technology, including
the research and development laboratories of the government
and of public/private sector industry, which are the main
sources of technological innovation. Hence, the emphasis
given to institutions in the scheme (Fig. 2). By and large,
spontaneous non-formal innovation (as distinct from minor
testing, modification and adaptation) by the people and
extra-institutional groups is believed to make a negligible
contribution to the stream of technology generation.
Perhaps this is because most innovations today require
large inputs from the accumulated heritage of scientific
and engineering knowledge which unfortunately is only
channelled through the formal institutional process.
(Whether this virtual exclusion of the populace from the
innovative process should continue to be the case is
another matter.)

---

[10]Quite deliberately, the neutral word "wants" has been used
at this stage. The resolution of "wants" into "demands" and
"needs"is discussed later.

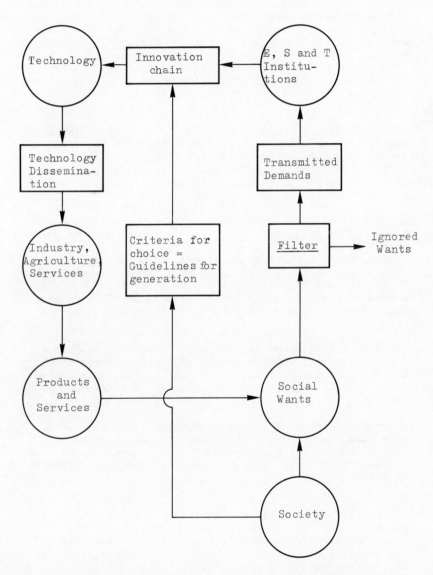

Fig. 2: <u>Generalized scheme for the development of technology</u>

(2) All social wants are not necessarily responded to by the institutions responsible for the generation of technology, viz., the educational, scientific and technological institutions. There is a process of _filtering_ these wants, so that only some of them are transmitted as _demands_ upon technological capability and the rest are bypassed by these institutions. In other words, there are _ignored wants_ which institutions do not seek to satisfy by research and development.

This filtering process is usually operated by decision-makers, firstly, in the bodies which control the research and development institutions, and secondly, within the institutions themselves. These decision-makers are either conscious agents of social and economic forces, or are unconsciously influenced by these forces.

In untempered market economies, _only wants which can be backed up by purchasing power_ become articulated as _demands_ upon the research and development institutions and the remaining wants are bypassed, however much they may correspond to the basic minimum needs of underprivileged people. Thus, like all commodities in these economies, technology too is a commodity catering to the demands of those who can purchase it, and ignoring those who cannot afford it.

(3) The generation of technology involves the so-called "innovation chain" which is the sequence of steps by which an idea or concept is converted into a product or process. This sequence of steps varies with the circumstances, but can often be schematically represented thus:

Formulation of research and development objective $\longrightarrow$ idea $\longrightarrow$ Research and Development $\longrightarrow$ Pilot-plant trial $\longrightarrow$ market survey $\longrightarrow$ Scale up $\longrightarrow$ Production/ product engineering $\longrightarrow$ Plan fabrication $\longrightarrow$ Product or process.

(4) It is essential to note that socio-economic constraints, and environmental considerations, if any, enter the process in an incipient form even at the stage of formulation of the research objective, and then loom over the chain at several stages. These constraints are in the form of guidelines or preferences or paradigms, for example, "Seek

economies of scale!"; "Facilitate centralised, mass production!" "Save labour!"; "Automate as much as possible!";
"Don't worry as much about capital and energy (in the days
before the energy crisis) as about productivity and growth!";
"Treat polluting effluents or emissions as externalities!",
etc. (These guidelines for generating technologies are only
another representation of the criteria for the choice of
technologies - guidelines stand in the same relation to the
generation process as criteria to the selection process).

Thus, every technology that emerges from the innovation
chain already has congealed into it the socio-economic
objectives and environmental considerations which decision-
makers and actors in the innovation chain introduced into
the process of generating that technology. It is in this
sense that technology can be considered to resemble genetic
material for it carries the code of the society which conceived and nurtured it, and, given a favourable milieu,
attempts to replicate that society.

(5) The technology that emerges from the innovation chain
will become an input, along with land, labour and capital,
to establish an industry or agriculture or a service, if and
only if the aforesaid socio-economic and environmental constraints are satisfied. Thus, it is not only the technical
efficiency of the technology, but also its consistency with
the socio-economic values of the society, which determine
whether a technology will be utilised.

(6) Social wants are not static. The products and services
that are produced create new social wants, and in this
process, the manipulation of wants through advertising, for
example, plays a major role, and thus the spiral:

Social wants ⟶ Products/Services ⟶ New Social
Wants ⟶ ....

The widespread generation of appropriate technologies depends,
therefore, upon the fulfilment of three important conditions:

(a) A filter which transmits basic human needs, particularly
the needs of the neediest, (viz., the urban and rural poor),
to the technology-generating institutions;

(b) The introduction of a new set of guidelines into the innovation chain, a set which is consistent with the criteria of appropriateness; and

(c) The existence of the requisite technological capability (trained and competent personnel, laboratories, workshops, test facilities, etc.) to complete the innovation chain.

The crucial question, therefore, is to what extent these conditions are satisfied in the developing countries. An exploration of this question can begin by noting that a causal spiral of the type represented in Fig. 2 is too simplistic in many ways, but particularly with respect to the social homogeneity that it implies. In point of fact, almost every developing country is polarised into a dual society[11]; an elite consisting of the richest 10-20% of the population, which usually includes industrialists, business-men and feudal landlords, politicians, bureaucrats, rich peasants, professionals such as doctors, engineers and scientists, and the bulk of organised white-collar labour; and the poorest 80-90% most of whom live in the rural areas, and the remainder in urban slums. In other words, dual societies are characterised by islands of affluence amidst vast oceans of poverty. Thus, in effect, a developing country consists of two "societies", which may not be spatially isolated from each other, but are separated by a wide chasm of incomes, consumption patterns, attitudes and life styles.

At the same time, the elite of developing countries practice a philosophy best described thus: "all that is rural is bad, all that is urban is better and all that is foreign is best", which means that there is a strong influence of the life styles of the developed countries upon the life styles of the elite in the developing countries. Hence, the tech-nology-society interaction scheme of Fig. 2 must be elaborated.

A simple version of such an elaboration is shown in Fig. 3 (which is closely related to that proposed by Herrera)[12]

---

[11]Dual societies are not to be confused with dual economies.

[12]Amilcar O. Herrera: Scientific and Traditional Technologies in Developing Countries, Chapter 13; Martin Robertson, The Art of Anticipation, London, 1975.

It is necessary, however, to make a few comments about this schematic representation.

(1) Little significance must be attached to the sizes of the circles, though,

(a) in the case of the row: Society, the circles 1-1, 1-2 and 1-3 have been drawn very approximately according to the relative size of the populations; and

(b) in the case of the rows: Educational, Scientific and Technological Institutions, and Technology, the circles 3-1, 3-2, 4-1 and 4-2 have been drawn very approximately according to the relative magnitudes of the R and D expenditures.

(2) The problem of the urban poor in developing countries is indeed a serious problem, and a more visible one to city-dwellers and foreigners. Nevertheless, it must be mentioned that:

(a) in most countries, the number of urban poor is much smaller than that of the rural poor;

(b) being subject so much to the powerful demonstration effect of the life styles of the urban elite, the urban poor share to a considerable extent similar aspirations; and

(c) the survival of the urban poor in the slums of metropolises generates many infrastructural requirements (e.g. services such as shelter, water, sanitation) which, for reasons of population and housing density, generate demands for technologies similar to those for the elite.

(3) Whereas there is a tremendous overlap between the wants in developed countries and those of the elite in developing countries (cf. circles 2-1 and 2-2), it is a characteristic of a dual society that there is virtually no overlap between the wants of the elite and the rural poor (cf. circles 2-2 and 2-3). The wants of the elite tend to be modelled on the pattern of the developed countries, in contrast to the rural poor whose wants correspond to the very basic minimum needs

of food, shelter, clothing, health, employment, etc.

(4) In dual societies, the bulk of the decision-making is
in the hands of the elite who are, therefore, responsible
for the filtering process which selects some wants for
onward transmission as <u>demands</u> upon the educational,
scientific and technological institutions, and shelves
other wants. In most cases, this elitist filtering process
functions in such a way that:

(a) the wants of the elite are almost wholly transmitted
    as demands requiring technological answers; and

(b) the wants of the poor are largely ignored even though
    they are an expression of urgent basic needs.

Since it is the satisfaction of these basic needs which
constitutes the essence of development, it follows that
an elitist filtering process is incompatible with development.

(5) The demands of the elite are picked up by educational,
scientific and technological institutions through the agency
of industry in the developing country, and industry in
developed countries, both of which sense in these demands
a major market. It is important, however, to note that
industry in developing countries is of two categories:

(a) indigenous industry which derives its technology
    from the national, educational, scientific and
    technological institutions; and

(b) industry which may be owned by government, native
    entrepreneurs or multinational corporations (or by
    two or three of these in different ratios), but which
    is based on imported technology generated in the
    institutions of developed countries.

Between these two categories, the linkage of the demands
of the elite is very much stronger with the second category
of local industry, viz., that based on imported western
technology developed by the educational, scientific and
technological institutions of the developed countries. This
is why the strong linkage 2-2 $\longrightarrow$ 3-1 is shown with a

continuous line and the weak linkage 2-2 ———→ 3-2 with a
dashed line.

(6) The operation of the filtering process to block the
transmission of most of the wants of the poor, i.e., the
basic minimum needs of the majority of the poverty-
stricken population, from the educational, sientific and
technological institutions is emphasised by the absence
of a linkage between the circle 2-3 and either circle 3-1
or circle 3-2. Of course, the linkage is not zero - for
instance, when the poor suffer from communicable epidemic
diseases, the elite is also vulnerable, and such needs of
the poor are obviously responded to effectively. Thus,
the filtering process is not conducive to development, and
particularly to rural development which in most developing
countries must constitute a major aspect of the development
process.

(7) In the absence of institutions to develop technologies
to meet the needs of the rural poor, the latter have no
choice except to fall back on traditional technologies
based on the reservoir of empirical knowledge accumulated
through the centuries. (cf. the linkage 2-3 ———→ 4-3 and
4-3 ———→ 2-3). The urban poor are generally victims of
rural impoverishment who migrate to metropolitan slums.
As such they not only carry over some of their traditional
technologies, but they are also forced to depend on urban
technologies. In addition, they innovate with the waste
materials and garbage dumps of the urban elite.

(8) There is very strong linkage 3-1 ———→ 3-2 between the
educational, scientific and technological institutions of
developed countries and those in developing countries,
the latter being modelled very closely on those of the
former. In fact, these institutions in developing countries
derive their patterns for research and development, including its emerging ideas, trends and fashions, stream of inspiration, experimental techniques and instruments, criteria
of excellence and source of recognition, from the counterpart institutions in the developed countries.

(9) In the generation of technology, the educational,

scientific and technological institutions of developing
countries invariably start with imported western technology as a starting point and as a model, hence the
linkage 4-1 ⟶ 3-2. Thus, they emerge (linkage 3-2 ⟶
4-2) after a process of imitation, adaptation and innovation (the innovation, rarely!) with a technology which has
been described as <u>naturalised</u>, i.e., adapted western technology.

(10) The satisfaction of the demands of the elite is much
more through western technology (this strong linkage is
shown by a continuous arrow 4-1 ⟶ 2-2) than through
naturalised technology (this weak linkage is shown by the
dashed arrow 4-2 ⟶ 2-2).

The above discussion of the technology-society scheme in
developing countries leads to important conclusions regarding the problems associated with the generation of
appropriate technology in these countries.

Firstly, the characteristics of these dual societies are
such that the filters do not emphasise the transmission of
basic human needs, particularly the needs of the neediest
(the urban and rural poor), as demands upon the technology-
generating institutions. The magnitude of the R and D
funding for problems related to basic needs is usually a
clear indicator of this bias, for it is very often significantly less than that for problems related to defense,
to glamorous technologies and to those aspects of the
industrial, agricultural and services sectors devoted to
the demands of affluent elites. Even if this funding bias
did not exist, and even if these institutions made deliberate efforts to respond to the basic needs of the urban
and rural poor, there is a serious problem in the identification of these needs. This problem arises because the
areas in which the urban and rural poor live, i.e., the
slums and villages, are not virgin territories unconta-
minated with the demonstration effect of urban life styles.
So, it is not simply a question of asking slum-dwellers
and villagers what their needs are - such a "questionnaire"
approach only results in their demanding needs similar to
the urban elite. For example, villagers invariably ask for
urban-style houses with reinforced-concrete-construction
(RCC) roofs, even though they are well aware that the
thermal comfort of such "modern" houses is often less than

the traditional thatched-roof houses. This request arises
from their clear understanding of the defects of traditional thatched roofs, which catch fire, leak, harbour
rodents, insects, and reptiles, are susceptible to termite
attack, and require frequent replacement though thatching
material is often scarce. This understanding results in
their feeling a need for an alternative roofing material
without the disadvantages of thatch as they use it, but
the only alternative which they perceive is an RCC roof.
Their lack of awareness of the range of possible roofing
materials, i.e. of the technical options, becomes an important reason[13] for the discrepancy between felt needs
and perceived needs. The discrepancy is serious because
these perceived needs usually require expensive western or
naturalised technologies, and therefore cannot be satisfied
by capital-starved developing countries.

Secondly, the intellectual domination of the developed
countries over the educational, scientific and technological institutions in the developing countries leads, in
the latter, to a virtually unexamined and unquestioned
introduction of alien and inappropriate guidelines, preferences and paradigms into the innovation chain, for
example, the implicit faith in "economies of scale". Unfortunately, these guidelines are largely unexpressed and
unstated. In fact, the participants in technological
innovation are rarely conscious that they cannot avoid
using preferences. The net result of not revealing, exposing, and evaluating the guidelines used in the process
of technological innovation in (or for) developing countries
is that the participants in innovation fall back on the
preferences of the industrialised countries. But the factor
endowments of developing countries may be fundamentally
different from those of developed countries. Under these
circumstances, the transfer of all those preferences related
to factor endowments is incompatible with development. Besides, developed countries have largely satisfied the
elementary minimum needs for most of their populations,
hence their technology has been increasingly oriented

---

[13]There are social reasons, too - in stratified societies,
the material appartenances of the upper strata become
status symbols avidly sought after by the lower strata.

towards other objectives (mainly towards non-essential luxuries and military applications). In developing countries, however, the main preoccupation has to be with elementary minimum needs from which large segments of their population are disenfranchised. Thus, guidelines and preferences related to products and services must necessarily be different in developing countries.

Thirdly, there is the problem of the thrust of technological capability. The task of generating appropriate technology appears to be impeded (a) by the type of technological capability that developing countries have and are currently growing; and (b) by the nature of linkages that their educational, scientific and technological institutions have and are forging with domestic and foreign societies.

Thus, most developing countries have followed a standard approach of establishing universities, institutes of science and/or technology, technical institutes and industrial laboratories modelled on the corresponding institutions in the developed worlds, with even their staff emulating counterparts in the industrialised world. As for the institutional linkages, Fig. 3 shows that the strongest links are with the demands of the elite, with counterpart institutions in the developed world, and with western technology. Furthermore, because of the inevitable financial stringencies, these institutions - like naturalised technologies - become, at best, cheaper and cruder versions of the corresponding western institutions, and at worst, complete parodies of the latter.

The reasons for this predicament are discernible from the technology-society interaction scheme for developing countries (Fig. 3). This predicament is an inevitable consequence as long as institutional linkages with the needs of the urban and rural poor and with traditional technologies are virtually non-existent (see the absence of arrows interconnecting circles 3-2 to 2-3 and 4-3), and are very strong with elite demands, with institutions in the developed world and with western technology (see the arrows interconnecting circle 3-2 with circles 2-2, 3-1 and 4-1). The situation is worsened by the fact that most teachers, scientists, and engineers are drawn from, and/or become part of, an elite which, in the dual societies of developing countries, is virtually cut off from its countryside and its rural poor, as well as from its slums and urban poor.

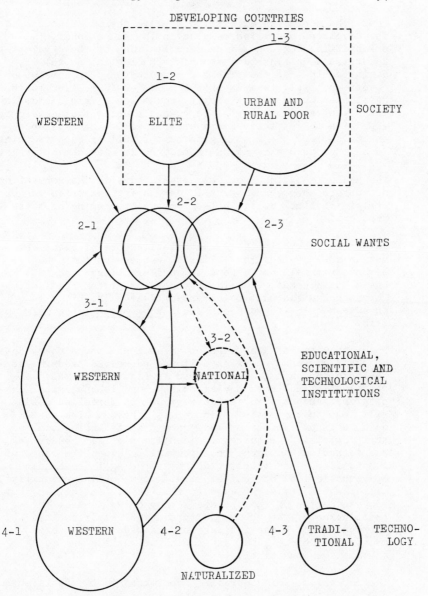

Fig. 3: Technology-society scheme for developing countries

Thus, the conscious attempt to emulate the institutions of
the developed countries results in the educational,
scientific and technological institutions of the developing
countries acquiring patterns of technological capability
distorted towards the problems preoccupying the developed
countries, i.e., towards problems largely unrelated to the
development needs of developing countries. This distortion
is only accentuated by the large-scale attempt of developing
countries to get their manpower trained in the developed
countries, for the most significant result of such training
is an increased alienation from domestic development tasks.

The analysis of the problems associated with the generation
of appropriate technologies in developing countries suggests
certain obvious prerequisites for the overcoming of these
problems.

The first prerequisite is the establishment of clear-cut
mechanisms to alter the filtering process so that the relevant institutions respond to, and are biased towards,
basic human needs, especially the needs of the neediest.
These mechanisms must be directed towards the creation of
an awareness of these needs and a commitment to satisfy
them.

The mechanisms of awareness creation must ensure the removal
of the discrepancy between felt and perceived needs. One
possible approach is represented in Fig. 4 from which it
can be seen that both social science and technological inputs are necessary so that a response to perceived needs
will lead to the satisfaction of felt needs. The social
science input, which may well come from technologists qua
sociologists, is required to identify the felt needs; and
the technological input is essential to widen the range of
perception of the target group by exposing it to a number
of technological options of varying cost and acceptability,
all of which satisfy the given felt need. It follows that
the generation of appropriate technologies to meet the
basic needs of the urban and rural poor requires, as a
prerequisite , close interactions between these target
groups on the one hand, and social scientists and technologists on the other.

The mechanisms for generating commitment to the satisfaction

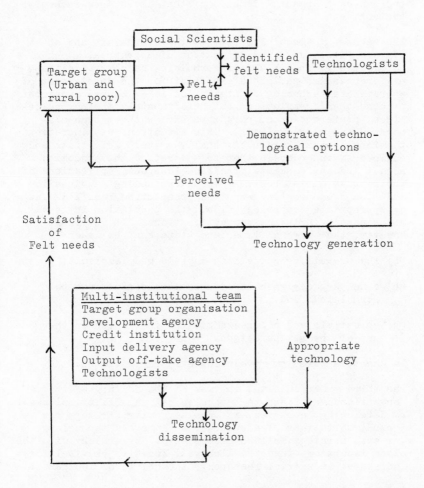

Fig. 4: <u>Model for dissemination of appropriate technologies</u>

of these basic needs must involve R and D funding policies
biased towards these needs, the creation within educational,
scientific and technological institutions of leaderships
and core groups which catalyse an increasing shift away from
the non-developmental demands of the elites, new incentive
systems, etc.

The second prerequisite is that the scientists and engineers involved in the generation of appropriate technology should absorb and/or formulate the new guidelines,
preferences and paradigms essential for the development of
appropriate technology. Absorption is of course an easier
process than formulation, but unfortunately the new guidelines, preferences and paradigms have not yet been elaborated and made explicit. This situation only increases
the intellectual burden on (challenge to?) scientists and
engineers, and compels them to understand the economic,
social and environmental ramifications and implications of
their areas of interest. This understanding requires in
addition close contact with the prospective beneficiaries
of appropriate technology, i.e., the population in rural
areas particularly the poorest sections. It is clear,
however, that several immediate steps must be taken:

(a) the prevailing guidelines must be made explicit;

(b) a new set designed to advance development must be
    formulated; and

(c) scientists and engineers must be instructed in the
    use of the new paradigms.

In short, a paradigm revolution must be initiated.

The third prerequisite, in view of the failure of the
conventional approach to technological capability followed
by developing countries, is an alternative strategy for
their institutions of education, science and technology.
One such strategy follows from Fig. 3. By analogy with the
close laboratory-industry link well-known to be vital for
successful industrial research, it consists of two parts:

(a) forging strong linkages between, on the one hand, the
    educational, scientific and technological institutions
    of these countries, and on the other hand, the needs of
    the urban and rural poor and their technologies; and

(b) drastically weakening the linkages of these institutions with elite demands and with institutions in the
developed world catering to these demands.

In practice, this alternative strategy is most effectively
implemented by each institution committing itself to a
neighbouring area[14], and to the generation of technologies
appropriate to the development of that area. It follows
from the understanding of development used in this conceptual analysis that a commitment to the development of a
particular area must mean a commitment to the needs of the
neediest, i.e., the urban or rural poor, in that area.
Further, since the generation of technological solutions
accessible and acceptable to the neediest is very often
likely to come through a transformation of traditional
technologies, a study and evaluation of these technologies
in the neighbouring area becomes an inevitable objective
of the commitment.

The alternative strategy needs simultaneous implementation
at a hierarchy of national, sub-regional and institutional
levels.

The national or macro approach should be directed towards
the preparation, on-going modification and refining, and
implementation of development-oriented technology plans,
in accordance with which the national R and D budget must
be framed and apportioned. This technology planning should
be linked with the process of selecting and choosing technology for development, so that better technological options
than those available can be identified, and research and
development work towards the development of these technologies can be initiated. Of course, insights into these
alternative technological options can only emerge from
greater sensitivity to the problems of the urban and rural
poor, and more intimate contact with their problems. Such
sensitivity to needs and intimacy with problems cannot be
attained by planning from the cloistered chambers of the
national capitals of developing countries. Hence, what is

---

[14]The size of this area can be a matter of convenience, it
may be a slum, a village, a cluster of villages, the poor
section of metropolises, a district or a province.

necessary is inputs from the grass-roots level, and herein
lies the importance of work at the institutional or micro-
level.

Virtually, all the educational, scientific and technological
institutions of developing countries (for instance, most of
those described earlier have, at least in an embryonic and
rudimentary form, the multi-disciplinary competence to
tackle the task of developing appropriate technologies for
meeting the basic needs of the needy. These functioning
institutions must, therefore, constitute the main basis
for the generation of appropriate technology through a
deliberate and formal commitment to the problems of development, and particularly to rural development. Such a
redeployment of efforts with existing infrastructures is
a far more effective measure than the creation of new
institutions for appropriate technology.

A basic assumption underlying this new strategy is that
institutional generation of technology will continue to
play a major, but hopefully not exclusive, role in innovation for development. It is envisaged that the actual
users and operators of appropriate technology, i.e., the
poor people themselves, will have a crucial role in innovation, particularly in the continuous testing, refinement,
and adaptation of new technologies. Indeed, it is hoped
that a constant interplay between institutional and popular
innovators will enhance the appropriateness of technologies.
What is rejected in the new strategy is the argument that
the institutions of education, science and technology in
most developing countries are so moribund and irredeemable
that only non-institutional voluntary groups can generate
the new technology. Such groups may have a part in technology generation (as distinct from technology dissemination), but it can only be marginal[15], in view of the inevitable limits to their multidisciplinary expertise and
their facilities for research and development. This
judgement rests, however, on the valid assumption that the
appropriate technology is not second-class technology and
that its generation is not a trivial exercise. For example,
innovative windmill design has been proved (cf. ASTRA's
experience) to require the same laws of aerodynamics as

---

[15] In contrast, their role in technology dissemination can
be, and often is, decisive.

that used in the design of jet-fighter wings; the dimension-
ing of biogas digestors needs as much chemical engineering
as the sizing of any chemical reactor; and an understanding
of the stresses in bullock-cart wheels requires the same
theory used for pre-stressed concrete.

The activities of institutions at the micro-level can be
linked together interactively at the sub-national or meso-
level, i.e., at the level of districts or groups of
districts corresponding to the next higher level of local
government. Here too the basic approach should be a
commitment to the problems of a particular area of the
country and to the problems of the poorest sections of the
people in that area.

This strategy of generating technological capability and
developing technologies through a commitment to a particular
area, i.e., to the problems of the poor in that area, is
likely to ensure satisfaction of the third prerequisite
for the generation of appropriate technologies, viz., the
growth of a new type of technological capability. In
addition, this same strategy is also likely to satisfy the
first two pre-requisites. Thus, the establishment of close
links between educational, scientific and technological
institutions on the one hand, and slums and villages and
urban and rural poor on the other, is the surest way of
creating awareness of the basic needs of the neediest. If
this awareness is transformed into a commitment to development, then the institutional and personal filters will start
responding to those social wants which correspond to the
basic needs of the poorest. Without such response the
generated technologies will not be compatible with development objectives. Similarly, it is only close contact with
rural areas and those below the poverty line which will
facilitate the formulation of new guidelines, preferences
and paradigms necessary for the generation of appropriate
technologies.

Dissemination of Appropriate Technologies

Turning from the technology-development process to the tech-
nology-dissemination process, it must first be noted that the
dissemination of conventional western technologies is a process the modalities of which have been established over
several decades. Further, the beneficiaries of these tech-

nologies are usually powerful and articulate groups ex-
pressing themselves through clear-cut market mechanisms.
As a result, commercial enterprises can, through profit-
seeking efforts alone, disseminate the technologies quite
successfully. In contrast, the dissemination of appropriate
technologies is a relatively more recent process and challenge. Also, its prospective beneficiaries are invariably
weak and inarticulate sections of society, e.g., the urban
and rural poor. These sections can rarely back up their
needs with purchasing power, i.e., they do not constitute
a significant market, and, therefore, the task of responding to their needs cannot be left solely to industry. Catalytic assistance from external sources is often essential
and inescapable. The purpose of this external assistance
should be to facilitate the technology implementation
process with technological know-how, with credit for equipment and working capital, with input deliveries and output
off-take, with managerial help and training programmes,
and entrepreneurial leadership. In addition, the beneficiaries, e.g., the urban and rural poor, must themselves
play an active role if the whole exercise is not to peter
out for the lack of popular participation.

It follows, therefore, that the dissemination of appropriate
technologies must be based on a multi-institutional effort
involving development agencies (either government or
voluntary agencies), R and D organisations, industry,
financial and credit institutions, input (e.g., raw
materials) delivery and product off-take (e.g., marketing)
organisations, management and personnel training institutions - and, of course, organisations of the beneficiaries
(e.g., cooperatives of the urban or rural poor).

This multi-institutional effort, which is so necessary for
the dissemination of appropriate technologies, implies that
a host of structures and procedures must be worked out for
each appropriate technology. In particular, attention must
be focussed on the procedures for the procurement of inputs
and credit, for the off-take of outputs, and for the management of organisations, training, manpower and finances. In
short, an entire hardware and software package must be
worked out in detail for each appropriate technology,
bearing in mind its specific features. Thus, the package
for appropriate road-building technology may be completely
different from that for mini-cement plants.

Too often, inadequate attention is directed towards the elaboration of these total packages, the general tendency being to assume that if the hardware (machinery, equipment or process) has been developed, the appropriate technology will diffuse under its own steam. In all except a few cases, even this hardware is rarely worked out with the same turn-key, engineered, finesse as the technologies of the industrialised countries - in short, the hardware development is rarely thorough. But even when this is the case, successful technology diffusion depends on the elaboration of the software. It is this shortcoming that has proved to be one of the major obstacles to the dissemination of appropriate technologies, and until this inadequacy is overcome, the process is unlikely to gain much momentum.

The insufficient emphasis on the development of the software aspects of appropriate technologies is what may be termed an _internal_ constraint on the successful diffusion of these technologies. In many circumstances, however, it is the _external_ constraints which are of far greater significance. Of these external constraints, the most important one arises from the fact that the partisan vested interests of the elites (or powerful groups within elites) in the dual societies of developing countries are often inimical to the adoption and diffusion of appropriate technologies. In such an unfavourable environment, inadequacies in the software aspects of these technologies are only amplified, and used against them in decision-making.

The above discussion of the dissemination of appropriate technologies shows that, though this process must be coupled with that of technology development, there are crucial differences between the two processes. Unfortunately, a blurring of these differences takes place too often, and it is therefore necessary to make them explicit.

Firstly, the _agents_ for the two processes are usually quite different - whereas R and D institutions (at the macro-, meso-, and micro-levels) are mainly responsible for technology generation, technology diffusion is usually the responsibility of a development agency, acting in coordination with the people, local self-government organs, R and D institutions, financial and credit institutions, and marketing organisations. Thus, technology generation can be achieved by the sole effort of R and D institutions, but technology diffusion must be a multi-institutional effort.

Secondly, the power structure need not necessarily be disturbed by the <u>generation</u> of technology, but it cannot but be affected by technology <u>diffusion</u>.

Thirdly, the levels of operation of the two processes are quite different - technology generation can be achieved at the <u>institutional</u> level; technology diffusion must be accomplished at the level of <u>society</u> (even a slum or a village is a mini-society).

Fourthly, and as a consequence of the above two differences, the technology generation process is much more <u>autonomous</u> than the technology diffusion process, in that, given (a) funds for R and D; (b) sufficient awareness and commitment among those doing R and D; and (c) the absence of direct political hostility towards the R and D, the generation of technology appropriate for development can be accomplished successfully.

In contrast, technology diffusion cannot be achieved against the wishes of the ruling groups in society. And, where the technologies to be diffused are against the vested interests of the privileged - which, in dual societies, they often are, if they are indeed technologies appropriate for weaker sections - then the success of the diffusion depends on the particular balance of power between various groups in society. The ruling group is rarely homogeneous, and if, within this group, some powerful sections, e.g., the urban elite, are not against the diffusion of appropriate technology for the rural poor, then the process stands a favourable chance. If, on the other hand, all the privileged sections are unitedly opposed to the technology, then the attempt to diffuse it is almost certain to fail; nevertheless the attempt must be made as an essential and integral component of the struggle of the under-privileged and its allies for a more just and equitable society. Thus, a necessary condition for the successful diffusion of technologies appropriate for the urban and rural poor is a large measure of active political support from the rulers of society.

Finally, the <u>role of the people</u> in the two processes is quite different. Though close <u>consultation</u> with the people is vital for obtaining better insights into felt needs, traditional solutions, local conditions, local materials and local skills, and though these insights are quite essential for ensuring the appropriateness of technology

(cf. Fig. 4), an R and D institution can in fact generate
technology without the active participation of the people
in the designs, calculations, experiments, fabrications,
etc. In other words, appropriate (including socially
acceptable) technology is unlikely to be generated by
R and D institutions without close consultations with the
people, but their active participation in the technology
generation per se is not necessary. This is not to deny
that widespread popular participation can raise the
efficiency and appropriateness of technological innovation
to a qualitatively higher level. Such popular participation should therefore be the objective, since an intimate
interplay between institutional and popular innovators is
an ideal state of affairs.

In contrast, the active participation and involvement of
the people is a necessary condition for technology diffusion.

These distinctions between technology generation and diffusion, particularly between social consultation and single-
institutional work for technology generation as distinct
from social participation and multi-institutional work for
technology diffusion, lead to some important perspectives
and conclusions, with regard to the role and scope for
appropriate technology institutions.

For instance, it is clear that institutions of education,
science and technology can assume - and successfully discharge - the responsibility of generating technologies.
If, however, these institutions also assume the responsibility for diffusion of technology, they must realise that:

(a) They will have to lead, coordinate and manage the con-
    certed action of a large number of institutions, viz.,
    development agencies, local self-government organs,
    financial and credit institutions, marketing outlets,
    etc., and

(b) They are almost sure to deviate from their charters
    of education, science and technology.

Whether they are structured and competent to discharge
this onerous responsibility is a moot question. In general,
it would be unwise for educational, scientific and technological institutions to assume this responsibility for
technology diffusion without being aware of all the implica-

tions and consequences. On the other hand, technology-
generating institutions must be an essential part of the
technology diffusion process - the vital need for their
active participation in the process follows logically
from the linkage between the technology generation and
diffusion processes (cf. Figs.1 and 4).

In conclusion, therefore, micro-level institutions of
education, science and technology can, and should, assume
leadership in technology generation, but not in technology
diffusion - they should only be members of the multi-
institutional teams to diffuse technology.

This conclusion need not be valid in the case of macro-
and meso-level institutions of science and technology, for
example, national and sub-national councils or departments
for science and technology, and institutions set up specifically for appropriate technology. Such institutions,
by their very nature and responsibilities, are already
removed from the laboratories, offices, drawing boards
and workshops where actual R and D work takes place, and
are, in fact, only promoting, catalysing and coordinating
R and D work. Hence, macro- and meso-level institutions
may be better equipped - compared to micro-level institutions - to lead the process of technology diffusion. But,
for such leadership to be effective, it should be formally
accepted by all the concerned institutions - development
agencies, local self-government bodies, financial institutions, etc. Even in the absence of such a formal acceptance
of the leadership of a macro- or meso-level institution of
science and technology, the latter two types of institutions
can still play a powerful catalytic role in the diffusion
of technology. Between them, if size and complexity of
organisation are in inverse measure of its speed and
effectiveness, it may turn out that macro-level institutions
are too ponderous and ineffective to play the catalytic role
efficiently. Hence, the hope for successful technology
diffusion lies in the hands of meso-level institutions,
e.g., the sub-national councils for science and technology.

Of course, the strengths of appropriate technology institutions operating at various levels can be made to
mutually reinforce each other ensuring that, on the one
hand, the macro- and meso-level institutions sponsor

technology-generation projects in micro-level institutions,
and on the other hand, the latter diffuse their technologies
through the higher-level institutions. With this perspective, it follows that micro-level appropriate technology
institutions can contribute to the dissemination of appropriate technologies by using the following mechanisms:

(a) Micro-diffusion in the specific region (slum, village,
    cluster of villages, etc.) to which they may be
    committed;

(b) Meso-diffusion by supplying appropriate technologies
    to meso-level institutions;

(c) Macro-diffusion through national, sub-regional and
    regional and international appropriate technology
    institutions;

(d) Long-term diffusion through educational programmes
    on the hardware and software aspects of appropriate
    technologies.

CRITERIA FOR ASSESSMENT OF GROUPS AND INSTITUTIONS

The above presentation of the factors determining the
capability of institutions to generate and disseminate
appropriate technologies serves as a basis for defining
the criteria to be used in the assessment of these institutions.

The definition of criteria can be taken to various levels
of detail, depending upon the ultimate purpose of the
criteria. Since the purpose here is not to assess the
capability of particular institutions, but to identify
the general problems facing a large number of micro-,
meso- and macro-level institutions in their task of
generating and disseminating appropriate technologies it
is obvious that the criteria must be neither too broad nor
too detailed. Further, the extent of disaggregation of the
criteria must be commensurate with the extent of information available about these institutions. The current lack
of detailed information indicates that the criteria need
only be disaggregated to an intermediate level.

Of course, there is considerable flexibility in this matter.

As the objective changes to an assessment of individual
institutions, and as more information is acquired (e.g.,
through questionnaires and/or actual visits); the criteria
can be made more detailed.

At the outset, three basic criteria regarding institutions
can be formulated with the aid of the following questions:

(a) Is it the purpose of the institution to develop
appropriate technologies, or to disseminate them,
or both develop and disseminate them?

(b) Does the institution operate at the micro-, meso- or
macro-level?

(c) Which basic needs - food, shelter, clothing, health,
education, employment, energy - is the institution
trying to satisfy with appropriate technologies?

If the objective of the institution is to develop appropriate
technology, then further criteria must be listed to assess
its capability with regard to this objective. These additional criteria must be generated from the three pre-
requisites for the development of appropriate technology
viz.,

(a) the institutional filter must select the basic needs
of the urban and rural poor and transform them into
demands upon its research and development capability;

(b) the institution must absorb or generate new paradigms
to guide its innovation chains towards the development
of appropriate technologies; and

(c) the institution must develop a new type of technological
competence and capability oriented towards satisfying
the basic needs of the urban and rural poor.

With regard to the operation of the institutional filter,
the criteria must be related to the institution's awareness
of and commitment to the problems of the urban and rural
poor. The importance of creating awareness leads to the
following criteria:

-    Does the institution have mechanisms for making contact
with the urban and rural poor?

- Are these mechanisms of the direct or indirect variety?

- Is the institution actually working in a slum/village/ poor part of a city/cluster of villages/district/ province?

The commitment of the institution to appropriate technologies is revealed by criteria such as:

- Is the avowed policy of the institution to emphasise the development of appropriate technologies?

- What percentage of the institution's funding goes towards appropriate technologies?

- Is the magnitude of funding, corresponding to this percentage, reasonably adequate for the technology development tasks it has undertaken?

- Has the institution created incentives (material and/or non-material) so that those of its personnel working on appropriate technologies concentrate wholly on the task, and those who are not working on these technologies turn increasingly towards them?

The acquisition (by absorption from external sources and/or generation from internal sources) of guidelines for innovation conducive to the development of appropriate technologies depends very largely on:

(a) Understanding the felt needs of the urban and rural poor; and

(b) Defining, through in-depth studies, the economic, social and environmental constraints which the appropriate technology must satisfy.

These requirements correspond to criteria such as:

- Does the institution possess in-house, the sociological expertise to define felt needs, or can it acquire this expertise by collaboration with outside institutions?

- Does it have the combination of economic, sociological and environmental expertise to define the various constraints on technological solutions?

-    Does it make explicit the constraints guiding its
     innovations?

The criteria pertaining to technological capability are
comparatively more straightforward and well-known. Nevertheless, it is worth stressing a few points regarding the
level of technological competence and capability required
for the generation of appropriate technology. These considerations arise from the fact that, invariably, appropriate technologies have been confused with "low" or
"primitive" technologies. This confusion arises because,
too often, the "advanced" character of a technology has
unfortunately been judged either by the trivial criterion
of "scale of production" or by the geographical origin of
the technology (anything from the developed countries is
ipso facto "high" or "advanced" technology), whereas in
fact it should be determined by the extent of the scientific and engineering thinking that goes into research and
development. An additional source of confusion is connected
with the question of simplicity. For use by the urban and
rural poor of developing countries, a technology (a product
or a process) may have to be very simple, but this does not
preclude the possibility of the R and D process (by which
the product or process is arrived at) from being ingenious
and subtle. To quote the adage "any fool can make a solution complicated; it takes a genius to make it simple!"

It must be admitted here that many appropriate technology
groups themselves have been responsible for initiating and
perpetuating the belief that the technical capability required for the generation of appropriate technologies is
of a lower order than that for western technology. (It
matters little whether this is due to inadequate understanding or to a deliberate attempt to bypass the established institutions of education, science and technology
in the developing countries.) The net result has been a
widespread belief that appropriate technology is "second
class" and not modern.

It is clear, however, from Figs. 2 and 3, and from the discussion of the filter and guidelines appropriate for development, that the only difference between western technology and appropriate technology is the difference in the
wants transmitted by the filter and in the set of preferences or paradigms guiding the innovation chain.
Otherwise, appropriate technology needs the same rigour

and thoroughness and must be developed from as sound a base
of fundamental science and basic engineering as is required
for western technology. In fact, most appropriate technology may need a much stronger foundation in fundamentals
because, after rejecting the technological paths well-
trodden by the developed countries, there is often no choice
other than going back to firm first principles.

The importance of this viewpoint must be seen in the context of hopefully well-meaning, but dangerous, advice that
developing countries should not invest in basic research.[16]
If such advice is acted upon, the result would be highly
detrimental to the development of alternative technologies.

Hence a basic condition for the generation of appropriate
technologies is that educational, scientific and technological institutions accept that appropriate technology is
as modern and advanced and sophisticated as western technology.

In the light of these comments, the criteria related to
technological capability are as follows:

-   Does the institution have an adequate infrastructure
    (laboratories, equipment, workshops, pilot-plant/test
    facilities, etc.) for carrying out the research and
    development work necessary for the generation of the
    hardware and software aspects of technology?

-   Does it have technical manpower with adequate training,
    expertise and experience?

-   Does it have the requisite information base to avoid
    unnecessary "reinventing the wheel"?

Even if the overwhelming emphasis of an institution is on
the generation of appropriate technologies, it is almost
certain to have some intentions of disseminating its
success. If so, there is a further criterion:

---

[16]Even though the share of basic research in the R and D
budget of most countries is rarely more than 5-10%.

-      Does the institution have any links with technology
       disseminating institutions?

Turning now to institutions whose primary thrust is the
dissemination of appropriate technology, several criteria
can be listed, such as:

-      Does the institution interact with the prospective
       beneficiaries in the definition of felt needs?

-      Does the institution know, or have access to, a
       sufficiently wide range of technological options?

-      Does it expose these options to the beneficiaries
       so that attempts to meet perceived needs will result
       in the satisfaction of the felt needs?

-      Is the institution part of a multi-institutional
       technology dissemination team?

-      Does the team include, or have ready access to:

       (a) a development agency?
       (b) technological expertise?
       (c) management expertise?
       (d) financial and credit institutions?
       (e) input procurement and product off-take organisations?
       (f) a beneficiary organisation?

In the technologies that the institution is disseminating,
does it work out total hardware-software packages, or does
it ensure that the team has worked out these packages?

In the case of institutions committed to the generation-cum-
dissemination of appropriate technologies, it is also
necessary to ask at what level - micro-,meso or macro-level
- is the institution participating in technology dissemination? Is it interacting with institutions which are
structured to achieve this dissemination?

The criteria proposed above can be used for the assessment
of the capability of national groups and institutions to
develop and/or disseminate appropriate technologies. However, they apply equally well to three other categories of
groups/institutions, viz., (a) regional/international institutions such as the International Rice Research Institute,

(b) groups in the developed countries such as the
Intermediate Technology Development Group seeking
to generate and diffuse appropriate technology for
the developing countries, and (c) groups/institutions
in the developing countries which run on the basis of
expatriates and foreign inputs. The question, therefore,
arises as to whether these three categories of groups/
institutions are related to the national capability of
developing countries.

A regional/international institution can either strengthen
national capability or weaken it depending upon whether it
works through and/or with national institutions, or in
competition with them. Very often, the vastly higher
salaries and the much better facilities in regional/
international institutions have the effect of undermining
the morale and confidence of national institutions,
particularly because the latter tend to lose some of
their best men to the former. Thus, a regional/international
institution is clearly not a part of national capability,
even of the country in which it is located.

The groups in the developed countries which seek to
promote appropriate technology in the developing countries
have done a great deal to spread the concept of appropriate
technology. However, many of these groups are funded by
the aid/foreign ministries/departments of the countries
in which they are located, and this fact has sometimes
created suspicions in the developing countries regarding
their intentions. These suspicions - irrespective of
whether they are justified or not - often result in
appropriate technology being viewed as a motivated recommendation from the developed countries and not as an
obvious conclusion from the predicament and circumstances
of developing countries. This view tends to be aggravated
by four tendencies of these foreign appropriate technology
organisations:

(a) they work very largely through personnel from
    developed countries who are paid much higher
    salaries than their local counterparts;

(b) they invariably bypass established institutions
    of education, science, and technology in the
    developing countries and work with volunteer and
    non-governmental groups even though the latter

often have less technical competence than the
former;

(c) with a few exceptions, the foreign appropriate
technology organisations have not built up local
information centres - in effect, they have retained control over appropriate technology
information;

(d) there are very few examples of these foreign
organisations initiating national efforts in
appropriate technology and then withdrawing
so that these efforts can grow in a self-reliant
fashion.

Hence, effort by appropriate technology organisations in
the developed countries must be viewed as international
action which is not part of national capability and must
be judged by the same criterion as other international
efforts, i.e., do they strengthen national capability?

Finally, there is the question of groups and institutions
which are located in the developing countries but which
have a large component of expatriates and foreign inputs.
Except in the few rare cases where these groups/institutions
are completely manned by expatriates and foreigners, these
groups must initially be considered part of the national
capability. It is important, of course, to determine at
an appropriate stage whether the expatriates are making
themselves redundant by generating native ability or
whether they are perpetuating their dominant position,
and whether the activity survives and grows after the
expatriates withdraw.

ASSESSMENT OF GROUPS AND INSTITUTIONS

The above criteria permit an assessment of the capability
of groups and institutions to develop and/or disseminate
appropriate technologies.

As a first step in this direction, a questionnaire (see
Appendix) was addressed to about 180 appropriate
technology organisations located in the developing

countries.[17]   The list included voluntary non-governmental
groups as well as established institutions. Of the
approximately 75 responses (i.e., about 40%) which have
been received, a detailed analysis[18] has been made of 54.
Of this sample, 27 (50%) orginated from Asia, 18 (33.3%)
from Africa, and 9 (16.7%) from Latin America. Based
upon these responses, and the impressions of a large number
of institutions/groups mentioned in the appropriate technology directories, a number of tentative assessments are
outlined below.

1. The large number of groups and institutions which
   have come up in the developing countries - most of
   them spontaneously - to develop and disseminate
   appropriate technology indicates the great deal
   of active interest in the concept. The formation
   of appropriate technology groups (e.g., DTC, TCC,
   ASTRA) within the framework of established institutions
   provides strong confirmation of a growing trend towards
   a new pattern of technologies.

2. The groups/institutions which have been analysed
   include all the categories: those that concentrate
   exclusively on developing appropriate technology
   (13%), those that disseminate it (11%), and those
   that are involved in development-cum-dissemination
   (76%). The fact that the last-mentioned category
   is predominant implies the widespread awareness of
   the importance of integrating the two processes of
   technology development and dissemination.

---

[17]The author also visited a few institutions, namely,
Technology Resource Centre, Manila; Institute of Small-
Scale Industries, Manila; International Rice Research
Institute, Los Baños; Asian Institute of Technology,
Bangkok; Korean Institute of Science and Technology, Seoul;
Technology Development Centre, Bandung; and the TOOL
Foundation, Amsterdam.

[18]The author wishes to express his gratitude to Miss
Lakshmi Reddy who rendered invaluable help with the
coding and tabulation of the responses.

3. Similarly, all levels of operation - local (17%),
   sub-national (22%), national (52%) and regional/
   international (5%) - are shown by the institutions
   in the sample. In many countries, however, it
   appears that appropriate technology organisations
   do not operate at the national macro-level, and
   consequently, the widespread dissemination of
   appropriate technologies may perhaps be inhibited.
   It is in this context that the proposed national
   centres, to be established with the assistance of
   the regional centres, have a crucial role to play.

4. It has emerged clearly that the term "appropriate
   technology" has different meanings to different
   institutions and groups. To almost half the analysed
   sample, it means appropriate to the area in which the
   group or institution works; to about 10%, it means
   appropriate to the sector (e.g., industry) with which
   the group or institution is concerned; and only to
   about 40%, does it mean appropriate to the poorest
   sections of society. (Of this latter category, 80%
   are concerned with the rural poor and only 20% with
   the urban poor.) Not too often does the definition
   of appropriate technology include an emphasis on
   basic needs, starting from the needs of the neediest.

5. Further, the concept of appropriate technology is
   generally restricted to production practices; it
   rarely extends to the appropriateness of products.
   Once again, this is because of the lack of a basic-
   needs emphasis.

6. The sectoral emphasis of the appropriate technology
   institutions is as follows: industry - 38% (and
   within this category 70% concentrate on agro-processing
   industries); agriculture - 30%; health - 11%;
   education - 8%; transport, housing, etc. - 13%.
   Thus, agriculture and agro-based industries account
   for almost 70% of the activities of these institutions.

7. The activities of the institutions which have been
   surveyed seem to span the entire range of basic needs
   with 27% of their efforts being devoted to food, 11%
   to shelter, 5% to clothing, 8% to health, 10% to
   education, 21% to employment and 18% to energy.

However, there seems to be a comparatively large
emphasis on alternative energy sources and devices,
particularly windmills. Appropriate energy technologies are of course vital, and if this emphasis
has been derived from a scrutiny of the felt needs
of the urban and rural poor, then it is quite
justified. Unfortunately, it is difficult to avoid
the suspicion that the emphasis is a carry-over from
the interests of appropriate technology groups in the
developed countries, and a result of the presence of
a large number of expatriates working on appropriate
technology in the developing countries.

8. Only about half the groups and institutions have
    direct contact with the main target groups. Almost
    80% of them have directed their activities towards
    the rural poor, in contrast, the urban poor are the
    concern of 20% of the groups/institutions. A mere
    2% are doing field work in slums, in comparison with
    64% in villages, and 34% in districts and provinces.

    It seems that most small groups (many of which are
    voluntary and manned by expatriates) are in direct
    contact with the urban and rural poor - hence, the
    strength of these groups in technology dissemination.
    In contrast, very few of the larger establishments of
    education, science and technology have this vital
    contact, for example, in the form of field stations
    or extension centres - this shortcoming is responsible
    for the relatively poorer performance of these insti-
    tutions at technology dissemination. It is note-
    worthy, however, that this shortcoming is being
    realised and that some of the most important institu-
    tional groups like DTC in Bandung are already
    operating field stations.

9. Even when large institutions of education, science and
    technology have declared interests in appropriate
    technology,[19] it appears that only a small percentage

---

[19] Of the groups/institutions which have been analysed,
about 12% (which is not insignificant) have emphasised
that appropriate technology is not the policy of their
parent organisations.

of their funding (less than 5-10%) goes towards this objective.

10. When established institutions of education, science and technology in the developing countries turn, even partly, towards appropriate technology, they are able, even with a short gestation time of 2-3 years to make significant contributions to technology development (cf., DTC, TCC, ASTRA). This only confirms the view that there is an immense potential and capability for appropriate technology generation in the established institutions of developing countries.

11. On the average, a group/institution in the sample has completed 4 technology development projects, and is currently working on 11 such projects with a manpower complement of 13 technical personnel including 3 Ph.Ds. An average budget can be a highly misleading figure because a number of groups are working with $5000-10,000 per year, and others are running million-dollar operations. However, an indication of the order of magnitude is provided by the average figure of $230,000 per year per group/institution which was noted from 29 Asian and African groups. It seems that, for technology development, the manpower and funding seems to be totally inadequate in relation to the vast spectrum of technologies which need to be developed and the enormous magnitude of the tasks. In fact, 53% of the groups/institutions analysed stated that their funds were inadequate.

12. With regard to incentives for work on appropriate technology, there is a vast difference between voluntary organisations on the one hand, and the established institutions of education, science and technology on the other. Whereas the former are able to attract manpower with extraordinary commitment, most of the latter have necessarily to deal with personnel the majority of whom have come into the institutions without a significant commitment to appropriate technology. Further, the incentive systems (the system of professional rewards and recognition, the criteria of excellence, etc.) in these established institutions usually operate away

from appropriate technology.

13. A significant 43% of the groups/institutions do not
     have in-house social scientists, and a quarter of
     these institutions do not have social science ex-
     pertise available even through collaboration. Of
     the 57% of groups/institutions which do have social
     scientists, 87% of these are economists, 73% socio-
     logists, 27% anthropologists and 17% political
     scientists.

     Thus, very few of the institutions and organisations
     dealing with the development of appropriate tech-
     nology have the sociological expertise (either in-
     house or available through collaboration) to define
     the felt needs of the target groups, e.g., the urban
     and rural poor. Even when there is an appreciation
     of the difference between felt and perceived needs,
     the identification of the felt needs is invariably
     done by scientists and technologists working as
     amateur sociologists.

14. A significant 26% of the groups/institutions admitted
     that they do not have the economic and sociological
     expertise to define the constraints which must be
     satisfied by a technical solution. There is little
     need for such expertise in dealing with western
     technology, because the constraints invariably per-
     colate to the R and D laboratories and the tech-
     nologists in them through market forces, but in the
     case of appropriate technologies, the prospective
     beneficiaries cannot articulate their demands
     through the market, and therefore there is no alter-
     native to including the definition of these non-
     technical constraints as an integral part of the
     innovation process. In fact, the innovation process -
     its first step being the formulation of the R and D
     objective - cannot even commence until the constraints
     are specified.

15. Perhaps because of the situation described in items
     13. and 14. above, most appropriate technology in-
     stitutions do not make explicit the precise economic,
     social and environmental constraints they seek to
     satisfy in their innovations.

16. A large percentage (40%) of the groups/institutions
     stated that they do not have in-house laboratories
     and workshops. Whereas almost 50% claimed to have
     testing facilities, only about 30% possess pilot-
     plant facilities. Further, only 19% stated that
     their R and D infrastructure was adequate.

     With regard to the infrastructure (laboratories,
     equipment, workshops, pilot plant/test facilities,
     etc.) necessary for technology generation, there is
     usually a vast difference between the small, often
     voluntary, organisations and the large established
     institutions of education, science and technology.
     The former usually struggle along with totally in-
     adequate infrastructures; whereas the latter, in
     most developing countries, are sufficiently equipped
     for the task of generating appropriate technologies
     (however ill-equipped they may be for competing
     with western science and technology.)

17. The same disparity usually exists with regard to
     the training, expertise and experience of the
     technical manpower deployed on the tasks - as a
     rule, the best manpower of a developing country
     goes to its established institutions. This assess-
     ment must, however, be qualified in two ways.
     Firstly, the influx of expatriates into the genera-
     tion of appropriate technology in developing
     countries is often associated with an influx of
     technical competence. Secondly, appropriate tech-
     nology attracts a number of DIY (do-it-yourself)
     inventors, and in some situations, such practical
     men are more useful than the qualified experts
     whose theory-oriented training and status-
     consciousness (characteristic of dual societies)
     renders them unfit for down-to-earth tasks.

18. The question of information back-up is crucial
     to the whole task of developing appropriate
     technologies. Hence the responses of the sampled
     institutions on questions related to information
     are important. 14% of the groups/institutions
     stated that they do not have a library to support
     their appropriate technology generation. Further,
     72% of those with libraries felt that these
     facilities were inadequate. Almost 80% of these

groups/institutions felt a need for journals, 66%
for photocopies of particular papers, 62% for books,
47% for experts, and 40% for relevant biblio-
graphies. These needs are emphasised by the fact
that 96% of the groups/institutions stated that
they sought to expand their expertise through the
medium of written material and 86% through
correspondence.

It seems therefore that most non-institutional
groups and even many institutional groups are
handicapped by a tremendous lack of necessary
technical information on appropriate technologies.
Even when such information is available, it is so
poorly organised that it is invariably irretrievable.
The net result is that acquisition of information
is very much a random affair - the result of the
grapevine, hearsay and chance contracts. This
state of affairs is only aggravated by the fact
that the available information systems are pre-
dominantly biased towards western technologies,
and are located in the developed countries.

19. Appropriate technology information is not only
     embodied in literature, it also resides in re-
     source-persons. It is no surprise, therefore,
     that 86% of the groups seek information through
     visits to persons working in the field, and 55%
     used visiting experts as a source of expertise.

20. In the case of technology dissemination, about
     35% of the groups/institutions which were studied
     are concerned with the local level and have
     personnel living in villages. The remaining 65%
     are involved with national and sub-national
     operations.

21. With regard to exposing technological options to
     beneficiaries so that their perceived needs become
     realistic and feasible, about 25% of the groups/
     institutions always adopt the procedure of de-
     monstrating options, and 50% follow this approach
     sometimes. But, 25% of the groups/institutions do
     not expose the various options. Thus, it seems
     that many groups and institutions get "hung up"
     on a few "pet" solutions, without exploring the full

gamut of possibilities. This is particularly the
case with institutions which are also generating
technologies, for they tend to become attached to,
and developed "vested interests" in, the solutions
which they themselves have generated.

22.   About 75% of the groups/institutions lead the
technology dissemination process, and three-
quarters of those which do not assume this leadership, form part of multi-institutional teams. In
doing the latter, 70% of the groups work with development agencies, 61% with beneficiary organisations, 48% with training organisations, 39% with
industry, 32% with credit organisations, 23% with
management organisations, and 18% with raw material
procurement and marketing organisations.

23.   About 90% of the groups/institutions devoted
attention to training of personnel, 74% to raw
materials procurement, 71% to organisational procedures, 63% to financial procedures, and about
50% to marketing. Thus, less attention is directed
to all software aspects of technology dissemination
than to the hardware aspects (equipment, process,
etc.)

The preliminary assessment which has been made above of
the appropriate technology efforts of groups and institutions in developing countries has revealed that there is
national capability in the developing countries for the
generation and diffusion of appropriate technology, but
this capability is in the incipient stages of growth. Besides, it suffers from a number of critical shortcomings
and limitations which need to be overcome.

1.   Even though a large number of institutions of education, science and technology have been established in the
developing countries, the vast majority of them do not
participate in the task of developing and disseminating
appropriate technologies for the urban and rural poor.
This non-involvement of established institutions in
appropriate technology represents a mis-orientation of a
valuable infrastructure which, if harnessed for appropriate
technology activities, could make contributions of orders
of magnitude greater than those currently being made
towards these ends.

2. There are many factors responsible for this non-involvement of established institutions, including their preoccupation with the thrust of the technologies of the developed countries. But, one crucial factor is the absence of national technology policies which favour the development and dissemination of appropriate technology.

3. The appropriate technology effort, as judged by the number of projects, the extent of funding, the development of manpower, etc., is quite inadequate compared to the enormous magnitude of the task, as indicated by the vast possibilities of appropriate technology and by the growing size of the population. This inadequacy extends over several aspects of the generation and diffusion of appropriate technology.

4. In the research, as distinct from the development, phase of the generation of technology, there are critical lacunae in the research infrastructures (laboratories, equipment, trained personnel, etc.) of groups which do undertake research, and in the linkages with research institutions of those groups which are not involved in research.

5. In the development, as distinct from the research phase of the generation of technology, there are limitations in the activities necessary to prepare the delivery of the technology on a turn-key basis. These limitations arise not only in pilot-plant trials, production engineering, etc., but also in the elaboration of all the software necessary for the technology to work, viz., raw materials procurement, organisational and financial procedures, credit, product marketing, training of personnel, etc. In other words, there is inadequate emphasis on total hardware-software packages for each technology.

6. The intrinsic appropriateness of generated technologies is partly limited by the absence of sufficient emphasis on a basic needs approach to the concept of appropriate technology.

7. It is also limited by the absence of sufficient inputs from the social sciences in the identification of felt needs, of various (economic and social) constraints on technical solutions, etc.

8. Yet another limitation arises from the inadequacy of direct contacts with the prospective beneficiaries.

9. The effectiveness of technology generation efforts is also handicapped by the information back-up required by appropriate technology groups and institutions.

In this matter of information, there are shortcomings not only in the organisation of the storage and retrieval of information, but also in its acquisition and dissemination. The acquisition problem is accentuated by one or more factors including its scarcity, cost and location (most appropriate technology information is situated in the developed countries). The dissemination problem is aggravated by the non-availability of reprographic facilities.

10. A crucial aspect of information flow is via people, and the difficulties of visiting active centres of appropriate technology, and of funding of the visits of experts and resource persons, constitute an important limitation of groups and institutions.

11. Such an exchange of personnel is particularly important for the dissemination of appropriate technology where the direct transfer of skills from experts to novices is often far more effective than the indirect transfer via documents. Unfortunately, finances for the funding of such exchanges of personnel, particularly between developing countries, is generally lacking.

12. Also lacking are field stations and extension centres and the corps of field and extension workers necessary to make such stations and centres effective.

13. The success of appropriate technology dissemination efforts depends to a large extent on the parallel functioning of such efforts at national (macro), state/province (meso) and local (micro) levels. The aim is to achieve a mutual reinforcing of efforts, the higher levels providing a favourable climate for the operation of the lower levels and the experience of the lower levels enlightening the perspective of the higher levels. A major shortcoming of current appropriate technology activities is the absence of appropriate technology groups and institutions at all levels-local, sub-national and national-or when such

groups/institutions exist, poor coordination of their
efforts.

14. There also seems to be an imbalance in the coordination
of appropriate technology efforts devoted to basic needs.
In particular, there seems to be insufficient concentration
on appropriate technologies specifically directed towards
employment generation.

15. Finally, the linkage between the two sections of the
appropriate technology movement, i.e., the established
institutions of education, science and technology on the
one hand, and the voluntary groups/non-governmental
organisations on the other, is too weak even though their
relative strong points are complementary.

CONCLUSIONS

The above discussion of the shortcomings and limitations
of groups and institutions currently engaged in appropriate
technology and potentially capable of contributing to such
work leads to the following questions:

(a) Is there a need for international/global action in
    the growth of national capability in appropriate
    technology?

(b) If so, what is the role for international/global
    action?

(c) What are the general and specific recommendations
    for such action?

The answer to the first question hinges upon an assessment
of whether every developing country can spontaneously, and
on its own, grow its national capability in appropriate
technology. The present situation is that there are a few
developing countries which have the scientific and technological infrastructures "to go it alone". Even in the
case of these countries, it is fairly clear that they
have much to gain by interaction with similar efforts in
other countries. Most countries, however, would be bene-
fitted greatly if their endogenous efforts are supple-
mented by external assistance and inputs. Bilateral and
multilateral relations can provide the required external

assistance and inputs, but these relations are very often
bedevilled by problems of mutual suspicion, of maintaining
equality and of differences on issues other than the
common objective of appropriate technology. In contrast,
international/global action is usually viewed favourably
by most countries. Thus, the need for international/global
action in building national capability in appropriate
technology is based on two factors:

(a) national efforts can be enhanced with external help;

(b) the most favourably viewed form of external help is
from international/global sources.

The role for international/global action must be derived
from the short-comings and limitations of groups and institutions currently engaged in appropriate technology
and potentially capable of contributing to such work. The
basic guideline for elaborating this role is that inter-
national/global action should concentrate on helping
national groups and institutions to overcome their shortcomings and limitations. In other words, the role of
international/global action should be primarily supportive
and catalytic. In playing this role, it should help the
activities of groups and institutions currently engaged in
appropriate technology work, as well as stimulate efforts
by groups and institutions which are not presently involved but which have the potential for making major
contributions.

Thus, international/global action should initiate policies,
programmes and projects to strengthen current efforts in
developing countries directed towards the development and
dissemination of appropriate technologies, and generate new
and additional efforts directed towards this same objective.
These policies, programmes and projects should include
several specific components which are discussed below:

1. Continuous efforts must be devoted to clarification
of the conceptual framework of appropriate technology,
particularly the introduction and propagation of a basic
needs interpretation of the concept.

2. Special policies, programmes and projects must be designed to catalyse the commitment of the established institutions of education, science and technology (i.e.,

institutes of science and technology, universities, poly-
technics, colleges, etc.) to the development and dissemination of appropriate technology. The success of this catalytic effort will depend, _inter alia_, upon the creation of
national and international climates which are favourable
to appropriate technology, as well as upon funding specifically directed towards appropriate technologies.

3. Substantial funding must be made available for current
and new appropriate technology efforts so that the magnitude
of the effort (number of projects, deployment of manpower,
etc.) is commensurate with the enormous task.

4. The highest priority for the development and dissemination of appropriate technologies which are specifically
directed towards employment generation _in rural areas_
(because that is where the largest numbers of unemployed
and underemployed are) and _in relation to the agricultural
sector_ (because that is the sector with which the bulk of
the rural population is associated). It follows that
employment-generation efforts must concentrate on:

(a) _water management_ including water harvesting, storage,
    lifting and irrigation to increase the number of
    harvests per year and the yields per harvest;

(b) _agro-processing industries_ to ensure that as much
    processing of agricultural products, by-products and
    residues takes place at or near the sites of agricul-
    tural production;

(c) _decentralised energy production_ (preferably based on
    renewable energy sources) to power those agricultural
    and agro-processing operations for which the use of
    inanimate energy is appropriate.

5. In the funding of particular programmes and projects,
the emphasis must be on support for field workers, extension workers, field stations, extension centres, pilot-
plant facilities, exchanges of personnel and experts
between groups, pilot projects, R and D back-up, etc.

6. Support must be extended for the working out of
complete hardware-software packages for each technology.

7. An appropriate technology information system must be

worked out as rapidly as possible for the acquisition, storage, retrieval and dissemination of information on appropriate technologies. This system must not be centralized in the developed countries. Instead, it must be based on a network approach involving linkages between national, sub-national and local information centres in the developing countries, and must be accessible to all groups and institutions working in the field.

8. Provision must be made for increasing contributions from social scientists (of various disciplines including social anthropology) to the generation and diffusion of appropriate technologies.

9. The establishment, if necessary, in each developing country of appropriate technology groups and institutions at the national, sub-national and local levels, must be facilitated, and the coordination of the efforts at these various levels must be promoted.

10. The Regional Centres for Technology Transfer must be stimulated to make appropriate technology their main focus.

11. The accent must be placed on appropriate technology in the new emphasis on Technical Cooperation among Developing Countries.

Questionnaire

(Strike out where inapplicable and put a ✓ mark in
appropriate boxes)

1. Name of Institution/Group within institution:

   Year of formation: 19

   Name of Director/Convener:

   Full mailing/postal address:

2. Name organisation(s)/agency or agencies/institution(s)
   which sponsored your institution/group within
   institution?

3. What is your area of operation? National ☐,
   sub-national ☐, local ☐

4. What is the main thrust of your activities?
   Technology Development ☐, Technology
   Dissemination ☐, Technology Development and
   Dissemination ☐

131

5. According to your understanding, does appropriate technology mean:

appropriate for the needs of the <u>area</u> in which you are working ☐?

appropriate for a particular sector of the economy ☐?

Industry ☐, Agriculture ☐, Services ☐?

appropriate for the poorest sections of society ☐?

Urban poor ☐, Rural poor ☐?

Questions pertaining to Technology Development

6. For which sector are you developing appropriate technologies - Industry ☐, and in particular Electronics ☐, Metallurgical ☐, Agro-based ☐, Consumer Goods ☐; Agriculture ☐; Health ☐, Education ☐, Transport ☐; Other ☐ (Please name)

7. For which basic needs do you consider your technologies appropriate: Food ☐, Shelter ☐, Clothing ☐, Health ☐, Education ☐, Employment ☐, Energy ☐

8. Does your group/institution have direct/indirect contact with the urban poor/rural poor?

9. Are you doing field work in: a slum ☐, slums ☐, a village ☐, villages ☐, district ☐, province ☐?

10. Is appropriate technology a policy of the institution to which you are attached? Yes ⟋⟋ No ⟋⟋

11. How many appropriate technology development projects have you completed? ⟋⟋
are you currently working on? ⟋⟋ (please insert number)

12. If the list of these projects is given in a report, please send report; otherwise please list titles.

13. What is the annual budget for all these current projects (only approximate figures need be given)?

14. What percentage of your institution's total annual budget is set apart for appropriate technology projects? ⟋⟋ (please insert number)

15. Do you consider your funds for appropriate technology projects adequate? Yes ⟋⟋ No ⟋⟋

16. Does your institution provide special incentives for those working on appropriate technology projects, for example, by way of higher salaries ⟋⟋, faster promotions ⟋⟋, greater R and D funds ⟋⟋, greater prestige ⟋⟋? (Please mention any other incentive scheme)

17. Does your group/institution include social scientists?
    Economists ⬭, Sociologists ⬭,
    Anthropologists ⬭, Political Scientists ⬭

18. Do you collaborate with social scientists from other
    institutions? Yes ⬭   No ⬭

19. Do you have social science expertise to define the
    economic, social and political constraints which
    your appropriate technologies must satisfy?
    Yes ⬭   No ⬭

20. Do you make these constraints explicit?
    Yes ⬭   No ⬭

21. Do you have <u>in-house</u> laboratories ⬭,
    Equipment ⬭, Workshops ⬭,
    Do you consider these facilities adequate?
    Yes ⬭   No ⬭
    Pilot-plant facilities ⬭, test facilities ⬭
    If not, do you sponsor work? Yes ⬭   No ⬭
    If you sponsor work in other institutions, how many
    projects have you sponsored ⬭? are you currently
    sponsoring ⬭?

22. Do the institutions in which you sponsor R and D
    have adequate facilities? Yes ⬭   No ⬭

23. How many research scientists and engineers are
    involved in your appropriate technology development?
    In-house ⬭? Sponsored Projects ⬭
    (Please insert number)

Of these, how many Ph.Ds ? ⟋⟋ (Please insert number)

24. Do you have a library for your appropriate technology
    development work? Yes ⟋⟋   No ⟋⟋
    Do you consider it adequate? Yes ⟋⟋   No ⟋⟋
    If not, what do you need? Books ⟋⟋, Journals ⟋⟋,
    Access to experts ⟋⟋, Bibliographies ⟋⟋,
    Photo-copies of particular papers ⟋⟋

25. Even though your main thrust is technology development,
    do you have links with technology disseminating
    institutions? Yes ⟋⟋   No ⟋⟋
    If so, are these links with industry ⟋⟋, government
    development agencies ⟋⟋, non-government development
    organisations ⟋⟋, farmers ⟋⟋? If links are with
    other types of institutions not in the list, please list.

Questions pertaining to Technology Dissemination

26. How many technology dissemination projects have you
    completed ⟋⟋? are you currently working on ⟋⟋ ?
    (Please insert numbers)

27. If the list of these projects is given in a report
    please send report; otherwise, please list titles.

28. What is your annual budget for your technology
    dissemination projects (only approximate figures
    need be given)?

29. If your primary objective is technology dissemination,
    do you have linkage(s) with technology development
    institution(s)? Yes ⬚   No ⬚.
    If yes, please name institution(s)

    If not, what is the source of the technologies which
    you disseminate - expertise gained by your staff
    before joining you ⬚, literature ⬚, visits to
    technology development institutions ⬚,
    correspondence ⬚, visiting consultants/experts ⬚

30. Do you attempt dissemination at the national ⬚,
    sub-national ⬚ or local ⬚ levels?

31. Who are the beneficiaries of your disseminated
    technologies? Industry ⬚, government ⬚,
    non-government organisations ⬚, urban poor ⬚,
    rural poor ⬚

32. Do you interact directly with the beneficiaries
    before identifying their felt needs? Yes ⬚ No ⬚

33. Do you expose the beneficiaries to more than one
    technological option before disseminating a particular
    option? Always ⬚, sometimes ⬚, so far not ⬚

34. How many personnel from your institution/group within

your institution live in the field (slum(s) ⁄‾⁊,
village(s) ⁄‾⁊, district ⁄‾⁊, province ⁄‾⁊)
where you are disseminating technologies?
Please state number ⁄‾⁊

35. Is your technology dissemination carried out under
     your leadership? Yes ⁄‾⁊   No ⁄‾⁊
     If not, are you part of a multi-institutional
     technology dissemination team? Yes ⁄‾⁊   No ⁄‾⁊

36. When you disseminate technologies, do you work with
     industry ⁄‾⁊, government development agencies ⁄‾⁊,
     non-government development organisations ⁄‾⁊,
     management experts ⁄‾⁊, financial/credit
     institutions ⁄‾⁊, input/raw materials procurement
     organisations ⁄‾⁊, product offtake/marketing
     organisations ⁄‾⁊, training institutions ⁄‾⁊,
     beneficiary organisations ⁄‾⁊

37. Before you disseminate technologies, do you ensure
     that, apart from the hardware (equipment, process,
     etc.), all the software aspects of the technology are
     worked out, i.e., raw materials/input procurement
     procedures ⁄‾⁊, organisational structure ⁄‾⁊,
     financial procedures ⁄‾⁊, training of personnel ⁄‾⁊,
     product off-take/marketing arrangements ⁄‾⁊?

38. Any other pertinent information.

Chapter 4
ACTIVITIES OF THE UN SYSTEM ON APPROPRIATE TECHNOLOGY

W.M. Floor

INTRODUCTION

A great variety of activities in the field of science and
technology are being undertaken within the UN system.
There are few subjects which are not covered by these
activities. This work is done by all agencies at different
levels throughout the world, ranging from preliminary
studies to actual surveys and the setting up of institutes
for R and D, as well as the actual execution of projects
with a common component of science and technology.[2]

A large number of collaboration agreements exist between
UN agencies with regard to problems of common concern, including reciprocal representation at meetings and the reciprocal right to propose items for the agenda of meetings.
Still coordination in the field of science and technology
within the UN system is badly needed. Apart from the
management problems caused by the existence of hundreds of
completely different activities, there is the question of
the autonomy and technical competence of the UN agencies.

The problem of autonomous development of activities within

---

[1]Member, Policy Planning Section for Development Cooperation, Ministry of Foreign Affairs, Netherlands.

[2]See for example, United Nations Economic and Social Council,
Institutional Arrangements for Science and Technology in the
UN system, E/C.8/29. ADD.I, 3 October 1975.

the various agencies with the resulting danger of overlapping, duplication, and unco-ordinated work is compounded
by the different constitutions and membership of the
agencies. An illustrative example is that of the ILO which
is empowered to engage in a wide range of social and
economic activities. Here overlapping occurs with agencies
such as FAO (rural development), UNESCO (training/education),
UNIDO (small-scale industries), WHO (workers' health
standards), and UNEP (housing). The same can be observed
with regard to the other agencies.

The fragmentation of science and technology activities is
further accentuated by the lack of coordination at the
regional and national levels. Such an unsystematic arrangement for assessing science and technology activities is
illustrated by the different positions taken by different
representatives of the same country in the various agencies.

No overall UN policy with regard to science and technology
yet exists. Furthermore, appropriate technology has only
recently received attention as a field of operation in its
own right. It is therefore not surprising that the UN
system lacks a suitable mechanism for ensuring a comprehensive perspective and focus for action on appropriate technology.

At present practically all R and D work takes place in the
advanced countries. And R and D in the field of appropriate technology is only a small fraction in comparison.
Under these circumstances, hardly any sizeable impact of
UN activities under status quo can be expected. However,
a unified system-wise science and technology policy,
accompanied by mutually complementary inter-agency programmes, could lead to some significant results at least
in a selected number of priority areas.

We start by examining the existing mechanisms which are
intended to play a coordinating role within the UN system.
We then review the on-going activities of the different
UN organisations which are engaged in appropriate technology. Better coordination possibilities through joint
planning of programmes relating to appropriate technology
are also discussed. The concluding section makes some
recommendations for improving the performance of existing
UN mechanisms. In particular, it examines the role of
UNDP in stimulating coordinated action at the country level.

SOME EXISTING COORDINATING MECHANISMS

Advisory Committee on the Application of Science and
Technology to Development (ACAST)

ACAST, being an advisory body, could be considered instrumental in coordinating UN activities in the field of
science and technology. Created in 1963 following the
United Nations Conference on the Application of Science
and Technology for the Benefit of the Less Developed Areas,
its major piece of work, the World Plan of Action (WPA) of
1971, only received a lukewarm welcome. Apart from the
list of R and D priorities and the special attention given
to appropriate industrial technology and product and plant
design, the WPA recommended the establishment of a fund,
as part of the resources to be made available to UNDP.
The fund was never established.

In response to General Assembly resolution 3168 (XXVIII),
and to Economic and Social Council (ECOSOC) resolution 1826
(LV) of August 1973, ACAST set up an Ad-Hoc Working Group
on Policy for Science and Technology, within the United
Nations system. Over the past few years the Group sub-
mitted various reports, most recently to the 24th Session
of ACAST in August 1978. This report[3] does not refer to
appropriate technology. It concentrates on the constraints
in harmonising policies, on alternative approaches to
harmonisation, on existing structures for coordination and
on the restructuring exercise for the Economic and Social
Sectors of the UN system. It concludes with a number of
alternative suggestions to improve institutional arrangements in the field of science and technology. The Advisory
Committee felt that the idea of a new Programme should be
emphasised without necessarily linking this alternative to

---

[3]Advisory Committee on the Application of Science and
Technology to Development; Report of the Ad-Hoc Working
Group on Policy for Science and Technology within the
U.N. System, E/AC.52/XXIV/CRP.2, 23 June 1978.

the creation of a special fund for science and technology
activities. The fate of this report and the very exercise
of formulating an overall science and technology policy for
the UN system, at this period in time, has become doubtful
both in view of the current UN restructuring process and
the preparations for the UN Conference on Science and
Technology for Development (UNCSTD) which is to take place
in August 1979.

ACAST, as requested by the Committee on Science and Technology for Development (CSTD), will submit this report to
the third session of the Preparatory Committee for the
Conference and also to all national focal points for the
UNCSTD, and regional commissions for their information.
ACAST's role has been important in promoting appropriate
technology. ACAST was among the first UN bodies to put it
on the policy makers desk by giving it some prominence in
the WPA. Later, ACAST advised the Second General Conference
of UNIDO (Lima, 1975) to do something about appropriate
technology.

More recently ACAST established an Ad-Hoc Working Group on
Appropriate Technology to review the work being done by the
various UN agencies and consider new developments and discuss possible contributions to the UNCSTD. In view of the
importance of the subject, however, one may well ask whether
such a temporary arrangement is sufficient.

At its 24th Session (August 1978) it was recommended that
ACAST prepare a paper on appropriate technology for policy
makers to be presented at UNCSTD and other related symposia
to be held in early 1979. The paper should clarify concepts and definitions of appropriate technology, present
the state of the art, review the work of the different UN
agencies and, to the extent possible, other activities
outside the UN system. Although ACAST is an independent
and objective body, it was agreed that the UN Office for
Science and Technology, in cooperation with interested
organisations of the UN system should prepare a first draft
of the paper in consultation with the Chairman of the
Working Group.

Committee on Science and Technology for Development (CSTD)

Apart from ACAST the UN has two organs which play an executive role in the field of science and technology within

the UN system. In 1971 the Economic and Social Council
(ECOSOC) decided to establish the Committee on Science and
Technology for Development (CSTD) "to eliminate any existing institutional gaps among the bodies and organisations
of the UN system dealing with specific science and technology problems".[4] Moreover, the CSTD was to provide policy
guidance and assist the ECOSOC in coordinating the
activities within the UN system in the field of science
and technology with a view to ensuring the utmost efficiency
and cooperation and avoiding duplication.

The CSTD activities until now fall rather short of expecta-
tions. It has not made any proposals for the improvement of coordination within the UN system. With regard to
the institutional arrangements for science and technology,
the CSTD did not even discuss this item during its third
and fourth sessions although the subject (and a position
paper) was on its agenda. This unfortunate development is
explained by several factors beyond the control of the CSTD
itself. CSTD is a political body set up to provide guidelines to the UN science and technology activities. However,
it has no subsidiary body to supply an analytical framework
to assess and evaluate the manifold UN science and technology activities. The Office of Science and Technology,
which acts as the secretariat of CSTD, has limited staff
although its workload has increased; it serves not only
CSTD but also ACAST and ACC Sub-Committee on Science and
Technology. It is no surprise then that CSTD has become
a deliberative organ dealing mainly with procedural,
administrative, and budgetary problems, instead of being
the most important political and executive UN organ in the
field of science and technology.

ACC Sub-Committee on Science and Technology

The Administrative Committee for Co-ordination (ACC) Sub-
Committee on Science and Technology is another UN organ
with a consultative task at the inter-agency level. Most

---

[4]ECOSOC Resolution 1621 (LI) of 30 July 1971, "Organisation
of the Work of the Council; Part B: Institutional Arrangements for Science and Technology".

collaboration agreements pertaining to more than two
agencies are concluded within the ACC and its subsidiary
bodies. The task of the Sub-Committee is to avoid duplication and conflicting programmes, to take mutually complementary action and in general to promote cooperation
and coordination where necessary. However, as experience
shows, consultation mostly takes place when programmes
have been adopted. The effectiveness of the Sub-Committee
is considerably hampered by the fact that no accepted conceptual framework exists, and each agency has its own list
of priorities. Moreover, the Sub-Committee is not equipped
to deal with specific projects but instead with general
principles and guidelines. Its mandate is very broad. It
does not really promote coordination of activities in the
field of appropriate technology.

REVIEW OF ONGOING UN ACTIVITIES

Many UN bodies are engaged, in one way or another, in
activities in the field of appropriate technology. To be
able to examine these activities objectively and critically
one needs a neutral and common yardstic; the more so since
"in a sense, all technologies may seem appropriate insofar
as at the moment of application, their choice may be economically rational".[5] Furthermore, the appropriateness of
technology has to be restricted to the possible use by developing countries. For although a certain technology may
be appropriate for a developed country it is not necessarily
at the same time appropriate for a developing country.

Opinions differ among the various UN agencies in respect of
the criteria determining the appropriateness of a technology. Some use more comprehensive definitions than others
with the danger that the term "appropriate technology" may
lose its relevance for a science and technology policy, let
alone for its critical review. This in itself is a reflec-
tion of the widely differing nomenclature for the term
appropriate technology such as: bare-foot technology, soft
technology, low-cost technology, village technology, inter-

---

[5]ACAST, Report of Ad-Hoc Working Group on Appropriate
Technology, p. 3 (E/AC.J2/XXIII/CRP.2, 10 June 1977).

mediate technology and many others. Each of these different
terms indicates the stress one wants to put on certain desirable aspects of the development and use of technology.
These terms also indicate that there is more than one concept of appropriate technology and that we have to choose
between the various options. It is not surprising, then,
to observe that most UN agencies view the appropriateness
of a technology as a relative concept, depending on the
development goals of the country and the use one wants to
make of such a technology. Logically and theoretically,
there is of course nothing wrong in taking such a view.
However, such an all embracing view of the concept tends to
lose relevance of the real development problems of the
developing countries. It runs the risk of losing its
meaning especially for the poor in those countries who are
the real goal of all appropriate technology development.

The development and use of technology therefore should be
geared to the satisfaction of the minimum requirements of
a family for private consumption such as adequate food,
shelter and clothing. Moreover, it should imply the satisfaction of community services such as safe drinking water,
sanitation, public transport, health and education. It
goes without saying that the satisfaction of basic needs is
of special importance to the rural and urban poor who constitute the majority of the population of the developing
countries.

An appropriate technology,therefore, should aim at:

(a) providing people with an adequate income through the
    creation of employment with low-cost per workplace;

(b) providing people with a (higher) income through better
    use of their resources, thereby raising productivity.
    This can be achieved by improving their production
    techniques as well as by providing them with better
    tools and equipment;

(c) providing people with goods and services commensurate
    with their needs and low income.

In the light of the above definition we examine below the
activities of various (though not all) UN organisations
relating to appropriate technology.

1. UNIDO

In 1966 UNIDO was created as an autonomous organisation
within the UN to play the central role and "be responsible
for reviewing and promoting the coordination of all activities of the UN system in the field of industrial development."[6]

Since its creation UNIDO has considered its main task to be
the most effective application of modern industrial methods
of production and the strengthening of the industrial infrastructure in the developing countries. This task entails
activities such as the dissemination of industrial knowledge,
and assistance to LDCs in solving technological problems,
especially in the adaptation and transfer of know-how, etc.

At its Eleventh Session (May-June 1977), the Industrial
Development Board approved a "Cooperative Programme of Action
on Appropriate Industrial Technology"[7] in response to a resolution which was adopted unanimously at UNIDO II, in Lima
(March 1975). UNIDO intended to mobilise the interest of
other agencies engaged in the field of appropriate technology. In fact, paragraph 32 of the document entitled
"Cooperative Programme on Appropriate Industrial Technology"
clearly states that it is a UN system-wide programme in
which other agencies will be invited to participate.

The programme is subdivided into four different fields of
action, namely:

(a) Promotion of technological research (paras. 42-60);

(b) Collection and dissemination of practical experience
    (paras. 61-66);

(c) Application of technology to rural development
    (paras. 67-80);

(d) Technologies for alternative sources of energy
    (paras. 81-84).

---

[6] UN General Assembly, Res. 2152 (XXI), para. 27.

[7] UNIDO, Cooperative Programme of Action on Appropriate
Industrial Technology, Report by the Executive Director,
ID/B/188, April 1977.

The programmes classed under (a) and (b) may not be primarily aimed at satisfying basic needs and alleviating poverty. They do aim, however, at making more efficient use of capital and natural resources. Many of the projects listed under (c), (paras. 67-84) may however be designed to satisfy basic needs. All of them explicitly aim at the provision of food, shelter and agricultural implements. They also aim at making more effective use of capital. It is not clear, however, whether these activities will lead to more employment. Moreover, some of these projects are concerned only with the dissemination of information (para. 80).

From the foregoing, it is evident that UNIDO is mainly concerned with the development and dissemination of technology which aims at cost-saving by making use of local natural resources and by improving the performance of industrial processes. Of course, there is nothing wrong in this; some UN body at least should be engaged in such an activity. However, most of these activities are not "appropriate" in the sense defined above.

This bias towards industrial development is also visible in UNIDO's establishment of an industrial and technological information bank which has to promote inter alia "the proper selection of advanced technologies."[8] This bank is regarded as an integral part of the implementation of the Cooperative Programme of Action.

One cannot escape the impression that UNIDO's current activities are not mainly centred on "appropriate technology for basic needs". Most of the activities are aimed at accelerating the transfer of advanced technology at better (and thus in a sense also appropriate) terms for the developing countries.

2. UNCTAD

UNCTAD was established in 1964 as an organ of the General Assembly. Its task is, inter alia, to accelerate and

---

[8] UN General Assembly, Establishment of an Industrial Technological Information Bank, A/32/116, Annex I, p. 2, 21 June 1977.

facilitate the transfer of technology from the developed
to the developing countries.

The main thrust of UNCTAD's activities is in the field of
the transfer of advanced technology. However since UNCTAD
IV, stress is also being laid on the appropriateness of
transferred technology. Since then, the transfer of technology is considered by UNCTAD as part of an integrated
policy aimed at strengthening the technological bargaining
power of LDCs.

Resolution 87 (IV) recommended that LDCs should:

(a) formulate a national technology plan to be an integral
    part of national development policy;

(b) establish national centres for the development and
    transfer of technology to implement, inter alia, the
    national policy;

(c) establish regional centres for the development and
    transfer of technology as a means of improving their
    negotiating strength vis-à-vis developed countries
    and of reducing their technological dependence.

UNCTAD has provided institutional arrangements to assist
LDCs in bringing about these policy recommendations. Apart
from studies regarding the international patent system, an
international code of conduct on transfer of technology,
and the experiences of several developing countries with
the transfer of technology, UNCTAD has advised countries
to formulate laws and establish institutions which will
improve their capability in dealing with issues of technological transfer. In addition, UNCTAD has started
training programmes for government officials in the field
of technology transfer. UNCTAD is also involved in the
establishment of regional and sectoral centres for the
development and transfer of technology in collaboration
with the UN regional economic commissions and other UN
organisations and specialised agencies.

Notwithstanding all these activities, one cannot classify
UNCTAD's work as primarily aimed at alleviating poverty
and satisfying basic needs. Since most of the technology
is transferred from the developed to the developing
countries, it is more often than not, inappropriate to the

latters' needs. It is evident that UNCTAD's activities, although necessary and useful, will, only in rare cases, contribute to the satisfaction of basic needs.

3. WORLD BANK

The World Bank was not engaged in promoting appropriate technology until 1971. That does not mean that all its activities prior to that date were inappropriate, but its policy then was primarily aimed at maximising efficiency and the growth of output. It was believed that the poor would also benefit from such a policy, by the indirect "trickle-down" effect. However, since 1971, the Bank has financed an increasing number of projects aimed at the alleviation of poverty.

The Bank considers a technology appropriate when it satisfies the following criteria: it is in accordance with the national development policy, the final product or service is useful and affordable by the consumers, the production process fits the socio-cultural setting and makes economic use of resources.[9]

The Bank policy with regard to LDCs is that the projects it helps to finance should be appropriate and that such projects should develop local capacity to plan for, select, design, implement, manage and, if necessary, adapt and develop appropriate technology.

The Bank's activities range over a wide variety of fields and include institution building, creation of infrastructure to support appropriate technologies, and economic and technical research. Most of these activities are aimed at the rural poor; activities such as the development of small-farm equipment, the introduction of new plant and crop varieties, improved agricultural extension services, credit mechanisms for the small farmer, educational facilities in rural areas, rural transport systems, promotion of rural industry, improvement of village water supply and rural road construction. In the case of the urban sector, the

---

[9]See World Bank, Appropriate Technology in World Bank Activities, July 1976, p. 5.

Bank promotes the efficient use of labour-intensive techniques in housing, urban water supply and waste disposal
projects.

A few years ago, the Bank launched a major research project on labour-capital substitution in civil construction
in developing countries. The initial (first) phase of the
study confirmed the technical feasibility of substituting
labour for equipment in road building with various design
standards. In the second phase of this project, field
studies were carried out in India and Indonesia on a
number of construction sites to gather data on labour and
equipment productivity rates obtained in the use of labour-
intensive methods. Until recently, the Bank maintained a
technology unit in Kenya to undertake research and development work in the construction of rural access roads with
an appropriate mix of labour, light equipment and tools.
Close collaboration was maintained by the World Bank team
with the ILO/UNDP team of civil engineers assisting the
Ministry of Works in the implementation of the Rural Access
Roads programme. These and other activities by the Bank
satisfy the criteria of our concept of appropriate technology on both counts: i.e. these activities lead to the
creation of employment and increase of income. They are
also capital-saving and aim at satisfying basic needs of
the poor.

4. ILO

The activities of the ILO in the field of appropriate technology are mainly executed by the Technology and Employment
Branch. ILO's technology policy has always been considered
to be an important part of its employment-oriented
development policy. The realisation that development
activities had not resulted in solving the employment
problem of LDCs led the ILO to launch the World Employment
Programme (WEP) in 1969. During the early period of the
WEP special attention was given to research projects in
different economic sectors. The commonly accepted platitude
that no real choice of techniques or of products existed
was proven wrong. A number of studies were published which
show the viability of alternative technologies as well as
demonstrate the relationship between technological choice
and employment generation and (to a lesser extent), income

distribution.[10]

Independent evaluations of WEP research on Technology and
Employment have been undertaken by the Swedish Agency for
Research Cooperation (SAREC); by Professor Thorbecke of
Cornell University, and by the WEP Research Evaluation
Meeting held at ILO headquarters in December 1976.[11] These
evaluations assessed the ILO technology research programme
in favourable terms and recommended more emphasis on the
institutional and socio-political bottlenecks in the
application of appropriate technology. Research now underway is increasingly concerned with these questions, in
particular, with the implications of government policies
for technological development, adaptations and innovations.

Apart from research on institutional and macro - as well as
micro -policy aspects of technology choice, development and
diffusion, the ILO is increasingly concerned with the implementation of appropriate technologies through technical
cooperation projects. The assistance given to the Kenya
Rural Access Roads Programme in collaboration with the World
Bank, and to the Guatemala Rural Roads Programme, are but
two examples of such technical assistance.

The first phase of the World Employment Programme (WEP)
culminated in the World Employment Conference (WEC) held in
June 1976. The WEC included an agenda item on "Technologies
for Productive Employment Generation in Developing Countries".

---

[10]See ILO, <u>Technology and Employment in Industry</u>, Geneva,
1975, <u>Employment, Technology and Development</u>, (Oxford University Press on behalf of the ILO), 1975; <u>Technologie et
emploi dans le commerce</u> (Librairie Droz, for the ILO, 1976);
<u>Employment and Technology Choice in Asian Agriculture</u>
(Praeger, New York, for the ILO, 1977); <u>Manual on the Planning of Labour Intensive Road Construction</u> (ILO, 1977);
<u>Technologies for Basic Needs</u> (ILO, Geneva 1977); and <u>Men
o r Machines</u> (ILO, Geneva, 1978).

[11]P. Thandika Mkandawire, <u>WEP Research: A Critical Review</u>,
Stockholm, September 1976; and Erik Thorbecke: <u>A Compre-
hensive Evaluation of the Research Component of the World
Employment Programme of the ILO</u>, 1976 (mimeo).

With the adoption of a basic needs strategy at the WEC, the ILO programme on appropriate technology is now being reoriented to ensure that the technologies that are appropriate for employment generation also assist in the satisfaction of basic needs of the poor.

The Programme of Action adopted at WEC called upon the ILO, in collaboration with other UN agencies to (a) help in the establishment of national and regional centres for the development and transfer of technology; (b) strengthen activities in the field of collection and dissemination of information on appropriate technologies; and (c) establish a Working Group on Appropriate Technology to ensure more concrete ILO action to implement the concept of appropriate technology in developing countries with the active support of its tripartite constituents.

Steps have already been taken to implement the recommendations of the World Employment Conference. The ILO's Programme and Budget for 1978-79 approved by the ILO Conference in June 1977, makes substantial provision for the dissemination of information on appropriate technologies. This work will be carried out in liaison with the Industrial and Technological Bank of UNIDO and with the proposed Inter-Agency Network on the Exchange of Technological Information. Some work on the preparation of "technical memoranda" which will consolidate technical and economic information relating to alternative technologies in particular branches of economic activity, has already started.

Along with UNCTAD and UNIDO, the ILO has contributed to the establishment of the regional centres for technology transfer and development. In order to promote inter-agency collaboration at the project formulation stage, the ILO invites UNCTAD and UNIDO to participate in its Steering Group on Technology and Employment.[12]

---

[12]Steering groups are informal and independent bodies which advise on WEP research. They consist of independent consultants drawn from research institutes and international agencies wherever appropriate.

To conclude, the ILO activites are aimed at the satisfaction of basic needs and the alleviation of poverty.
They show that an appropriate technology-mix - an optimal
combination of labour-intensive and capital-intensive
techniques - is the most viable proposition in a basic
needs strategy.

5. FAO

FAO was established in 1945 as a specialised agency of
the United Nations. Under Article I of its constitution,
the FAO's task, in relation to activities on appropriate
technology, was to promote and, where appropriate, recommend national and international action with respect
to:

(a) scientific, technological, social and economic
    research relating to nutrition, food and agriculture;

(b) the improvement of education and administration
    relating to nutrition, food and agriculture and
    the spread of public knowledge of nutritional and
    agricultural science and practice.

Activities on appropriate technology cover all the three
FAO Major Technical and Economic Programme areas. Under
the Agriculture Major Programme, FAO's stated objective
is to help in the development, transfer and local
adaptation of appropriate technology covering three main
fields: land and water resources and use, crop production
and livestock production. Under the Fisheries Major
Programme, proposed work includes the establishment or
strengthening of fishing technology services in Asia,
Africa and Latin America through programmes jointly
undertaken with national institutions.

Finally, the recent initiative taken by FAO to start a
post-harvest losses programme is an important contribution to the development and application of appropriate
technology for basic needs.

6. UN Economic Commissions

Although evidence may be incomplete, it would appear that
the UN Economic Commissions are not really actively
concerned with appropriate technology beyond producing
the well-meant but familiar and essentially simple
statements directed against the application of inappropriate technology.

There are nevertheless signs that this somewhat complacent
attitude is now subject to change particularly in the
case of the Economic and Social Commission for Asia and
the Pacific (ESCAP). A number of projects in the work
programme of ESCAP reflect an increased emphasis on
creating employment and meeting basic needs of the rural
poor[13] and related directly to technological aspects of
rural, small-scale industrialisation. ESCAP operates a
regional network for agricultural machinery, concerned
with design, development and commercialisation of agricultural implements and tools suited to meet the requirements of small farmers.

The Regional Centre for Technology Transfer, newly
established in Bangalore (India) under ESCAP auspices,[14]
should be expected to devote a major effort to advise
member countries on the choice of appropriate technology,
even though, oddly, no explicit reference of such a
function is made in the project document of the Centre.

Other projects are: a regional network for agro-industry
information, a project on bio-gas technology and utilisation

---

[13]See report of ESCAP XXXII Session, March/April 1976 as
included in ESCAP Annual Report (March 1975-April 1976)
presented to sixty-first session of ECOSOC as document
E/5786, July 1976.

[14]See Chapter 3.

and another on guidelines for the development of industrial
technology in the region.

Along the lines of the ESCAP Regional Centre, the Economic
Commission for Africa (ECA) also organised an inter-agency
mission and a ministerial Conference which recommended the
establishment of an African Regional Centre for Technology.
This Centre is expected to start its operations soon. As
in the case of the ESCAP report, the ECA report also avoids
an explicit reference to appropriate technology despite the
fact that the need for it is implied.

7.   UNDP

The UNDP, although essentially an agency to fund development
projects executed by specialised agencies, may exert considerable influence in financing projects with an appropriate
technology component. The UNDP philosophy concerning appropriate technology will largely determine the extent to which
concrete efforts are made by the UN system as a whole in
this area.

It is of interest to note that UNDP sponsored a meeting on
Appropriate Technologies in the Industrial Sector in
April 1975. Its Global and Interregional Programme for
1977-1981[15] states that "another key area for international
action is research and development aimed at strengthening
developing country capacities to develop and adapt technologies appropriate to their national objectives and
particular economic, social and employment situation". The
Programme calls for investigations into "the alternative
approaches that will more effectively attack mass poverty,
meet basic needs of all sections of the population and
integrate marginal groups into the development process".
It designates technology development and adaptation as a
"priority area which requires early investigation towards
possible UNDP activity during the 1977-1981 cycle".

Somewhat more reassuring is what the former UNDP Deputy
Administrator told the 1976 World Employment Conference:

---

[15] UNDP Governing Council, 23rd Session, January 1977, The
Global and Interregional Programme document DP/216 of
20 September 1976, para. 25.

"looking beyond, we are anxious to play a much more active role in the development and dissemination of appropriate technology".[16] Commenting on the proposed 'Consultative Group on Appropriate Technology' he expressed preparedness on behalf of UNDP "to get together with the ILO and our other agency partners as well as government partners to discuss the next step in providing a proper habitat and a home within the UN system for efforts to promote appropriate technology".

8. UNEP

On 15 December 1972, the United Nations General Assembly established UNEP and invited the organisations of the United Nations system to undertake coordinated programmes with regard to international environmental problems. Since UNEP activities cover areas of interest to other UN agencies, formal and informal relationships between UNEP and some of these agencies were established during the last two years.

In the technology field, UNEP has, up to now, been mainly concerned with developing criteria and methodologies for the identification and selection of what is termed by UNEP as "environmentally sound and socially appropriate technologies". UNEP has mainly produced policy papers dealing with parti-cular sectors (e.g. human settlements) or concepts (e.g. environmentally sound technologies). A few UNEP projects involved the promotion and dissemination of specific tech-nologies. Some of these projects include the establishment of rural energy centres, the publication of guidelines for the production of handpumps for rural areas, and the pro-vision of financial and technical aid to developing countries in the field of building technology.

This limited involvement in the field of appropriate tech-nology is understandable since UNEP tends generally to con-centrate on activities not covered by other UN agencies (e.g. sea water pollution, and depletion of the ozone layer). It is therefore expected that UNEP will provide support to

---

[16]See World Employment Conference, 4-17 June 1976, Summary Record No. 10, p. X/5.

activities of other UN agencies in the technology field
rather than undertake activities on its own.

9. WHO

Following the consideration of a report on "Health technology relating to primary health care and rural development" in May 1976, the twenty-ninth World Health Assembly
established a programme of Appropriate Technology for
Health (ATH). During 1977, the foundations were laid for
the first medium-term programme of ATH for 1978-1983. A
report by the WHO Director-General outlining the general
aim and specific objectives of the ATH programme was considered by the thirty-first World Health Assembly in May
1978 which, in a resolution, requested the Director-General
to intensify involvement of member States in the further
development of a global plan of action for the programme of
Appropriate Technology for Health.

The ultimate goal of the ATH programme is to promote national
self-reliance on problem-solving in primary health care delivery and to reduce the existing dependence of developing
countries on the industrialised countries for technological
support. Priority will, therefore, be given to measures
concerned with improving the health of underserved populations and to procedures which facilitate the decentralisation of services, production of goods and sharing of information.

The four principal functions of ATH programme are viewed as:
direct cooperation with countries, collaboration with
appropriate technology groups and with selected national
institutions, evaluation of progress, and dissemination of
information. It is emphasised that the new programme cannot
be restricted to the health sector, but must emerge as a
collaborative effort in which other sectors and agencies
are involved, and in which the appropriate technology groups
will play an important role.

Current WHO activities in appropriate technology include
quality improvements in community water supply in Indonesia
by research and development of bamboo as a piping material.
Special programmes exist for health laboratory technology,
basic radiological technology and immunisation technology.

The 1978-1983 medium-term programme for ATH[17] will focus
on a wide range of activities of a promotional nature including the dissemination of information. Already an ATH
Directory containing a list of institutions and persons
involved or interested in working in the ATH field has
been published. An updated and expanded version will be
published twice a year. Collaborative programmes of research for the development of new appropriate technologies
will also be increased.

*   *   *

It seems clear from the foregoing, that activities in the
field of science and technology by various UN agencies are
not necessarily concerned with appropriate technology. We
have based our conclusions in the light of the objectives
of employment generation and the satisfaction of basic
needs. There is no doubt that all UN activities in developing countries are ultimately aimed at improving living
conditions of the poor. However, many UN policy statements
do not progress beyond that general statement, with which
there can be but little dispute.

The poor have become a target group for the agencies concerned; yet only a few UN agencies care to identify certain
well-defined target groups with programmes geared to their
special needs. It is also very rarely argued why new programmes and policies are more promising than the past
policies. It is insufficient to conclude that the policies
which were in force until now did not benefit the poor
adequately and then decide to focus one's policy primarily
on the poor. This begs the question: can one view the poor
in isolation from the rest of society?

It is therefore not enough to gear projects specifically to
the needs of the poor. Such projects should also be relevant to the development policies adopted by the LDCs. The
choice of instruments, targets and technology should, therefore, be consistent with the over-all development policy.
It is only then that a certain technology is not only appropriate per se, but also fits appropriately in its larger development context. How difficult it is to implement such a

---

[17]"Appropriate Technology for Health"; Report by the
Director-General, Document A31/14 of 6 April 1978.
See especially Annex III.

policy is shown in this paper by citing the experience of
the various UN agencies. Only the World Bank states that
results can be very meagre given the existing socio-economic
structures in the LDCs. Appropriate technology can only be
effective and useful when the satisfaction of basic needs of
the poor as a policy is vigorously pursued by the national
governments concerned.

It would appear that most of UNIDO's activities are indirectly aimed at providing employment in the long run to
the grey mass of the unemployed urban poor. However, part
of its activities which are directly aimed at rural development may create additional employment. Yet it is not clear
who is to benefit from the extra employment thus generated.

A similar observation can be made with regard to the technology activites of UNEP and UNCTAD. There is no doubt
that UNCTAD's activities aim to strengthen the position of
the developing countries vis-à-vis the developed countries.
But these activities do not necessarily affect the LDC's
internal situation. Take for example, the case of the
centres for the development and transfer of technology.
One does not know who these centres are supposed to serve.
Unless the activities of these centres are carefully planned
and directed specifically at the poor target groups, it is
most likely that these centres will cater largely for upper-
income groups in the modern sector.

The few UNEP technology projects which have been started for
the benefit of the rural poor appear to have a strong bias
in favour of those who are already well-off. An exception
may be the UNEP hand-pump projects as these are public uti-
lities. In the long run, of course, UNEP's preoccupation
with subjects such as desertification and eco-development
may have a positive impact on the position of the poor.

The assault on world poverty announced by the World Bank is
less spectacular and serious than one is led to believe.
Although the World Bank has identified the 750 million poor
in the LDCs as its target group, in reality, its policy is
mainly aimed at the upper strata of the poor. It is true
that the Bank lending has been increasingly shifted to
agriculture and to the poorest countries. Yet most lending
activities are aimed at those countries which are less poor
than most and, in the poorest countries, to those who belong
to the "haves" among the poor.

The World Bank describes the following categories of the
rural poor: the small farmers, the tenants, the share-
croppers and the landless labourers. Most of its activities
are however aimed at the small farmers. The Bank envisages
the promotion of a minimum package to the small farmers
which will offer a rate of return at least equal to the
opportunity cost of capital. However, the Bank admits
that "the social and economic stratification in many South
Asian countries would seem to preclude widespread application of the minimum package approach there where the bulk
of the rural poor lives".[18]

The same bias towards the problems of the small farmers is
to be noted in the Bank's policy paper on agricultural
credit, which is now the largest component in the Bank's
agricultural lending. Here the main criterion used is the
productive capacity of the borrower's holding which excludes
such poor groups as tenants, share-croppers and landless
labourers. Those who are reached "constitute at best only
1% of the 100 million small farmers in the developing
world".[19]

ILO seems to have avoided the problem of identifying the
target groups for its appropriate technology programme. It
certainly does not look as if ILO is only aiming at satisfying basic needs of the bottom 40% of the poor. One gets
the impression that ILO's position is not very different
from that of the World Bank.

WHO is specifically concerned with health, one of the core
basic needs. Its programme, though relatively new, is directly addressed to the delivery of cheap health services
to the poor.

BETTER COORDINATION WITHIN THE SYSTEM

From the observations above it follows that some effort
should be made to coordinate better the modest UN activities

---

[18] World Bank: The Assault on World Poverty, Johns Hopkins,
1975, pp. 42,50.

[19] Agricultural Credit, A World Bank Paper, May 1975.

in the field of appropriate technology. Moreover, a common
UN strategy to the problem of science and technology in
general and of appropriate technology in particular is
highly desirable.

The lack of a conceptual framework makes the formulation of
coordinated UN programmes impossible. It is therefore surprising to observe the ritual repeated at each session of
the ACC Sub-Committee on Science and Technology when the
traditional item of the agenda, namely the World Plan of
Action (WPA) is discussed. One dutifully promises to de-
liver this or that supplement to the WPA. The relevance
of the WPA for the discussion under another item of the
agenda, namely, UN policy in the field of Science and Technology, strangely enough, is not mentioned in any ACC discussion. Hans Singer was quite right when he wrote that the
follow-up of the WPA is "a depressing spectacle, only some
subdued lip service has been paid, but no real attempt has
been made to follow through".[20]

The failure of WPA cannot be laid at the doorstep of ACAST.
The WPA was conceived in close cooperation with the agencies.
When the WPA was accepted and no funds were made available
by the UNDP to implement it, the agencies just ignored its
existence.

In 1976 the ACC Sub-Committee on Science and Technology put
the subject of joint planning on its agenda. All agencies
agreed "that some form of constructive coordination or joint
planning could be of positive value to many organisations".
The agencies hastened to add, however, that joint management
was not desirable. Participation in joint planning was
viewed as something voluntary and one was not required to
submit a report of activities if one did not wish to do so.

It rather looked as if no one was prepared to agree on
common key areas, and indicate possibilities for multi-
disciplinary and inter-sectoral cooperation. This impression
was strengthened by the lack of appreciation shown by most
agencies to an ILO proposal to have "outsiders" appraise the
programmes of agencies in a particular field. Such an out-

---

[20]Hans Singer, "Five Wasted Years", Focus, No. 1. 1976.

side appraisal could function as a temporary substitute for
the absence of a conceptual UN framework. However, the
prospect that the autonomy of the agencies would be curtailed
proved to be an effective antidote.[21]

Nevertheless, it proved possible for the ACC Sub-Committee,
during its 26th Session in November 1977, to agree upon a
first subject area for formulation of joint programmes and
projects. The Sub-Committee unanimously recommended for
this purpose, "technologies for low-cost construction" which
would include low-cost housing, storage facilities for food
and water, building materials and low-cost road construction.
It was decided that a working group consisting of organisations actively working in the above areas should be responsible for undertaking in-depth studies. The Office for
Science and Technology would arrange for the first meeting
for such a group some time in the middle of 1978".[22]

During the 28th Session (August 1978), of the Sub-Committee
this item was however taken off the agenda in order to give
the new UN Assistant Secretary-General for Programme
Planning and Coordination an opportunity first to examine
various efforts at joint planning before the matter is discussed at ACC.

CONCLUDING REMARKS

We have noted that the UN System has three organisations
dealing with the coordination of Science and Technology
activities: ACAST, an advisory body giving expert advice;
CSTD, a political body providing over-all direction, and
ACC Sub-Committee on Science and Technology, an inter-agency
body aiming at coordination. All these three bodies are

---

[21] ACC, Report of the Twenty-Fifth Session of the ACC Sub-
Committee on Science and Technology, 20-22 July 1977, doc.
COORDINATION/R.1227, pp. 4-6.

[22] ACC, Report of the Twenty-Sixth Session of the ACC Sub-
Committee on Science and Technology, 3-4 November 1977,
doc. COORDINATION/R. 1258, pp. 9-12.

serviced by the UN Office of Science and Technology. Since
all decisions concerning a UN Policy on Science and Technology are being deferred until after the UN Conference on
Science and Technology to be held in August 1979, we hope
that this World Conference will be able to formulate and
agree on a UN Plan of Action. This will ensure that more
harmonious and integrated programmes can be formulated and
carried out both within the UN system and at the country
and regional levels. If such a Plan of Action is formulated,
ACAST could undertake the task of providing expert advice on
a continuing basis. Moreover, ACAST should also continue to
advise the CSTD on institutional arrangements on science and
technology and on programmes of the UN agencies in this area.
Such a role by ACAST would no doubt increase the effectiveness of CSTD.

It would be desirable to convert the ACAST Ad-Hoc Group on
Appropriate Technology into a permanent one in order to
ensure that appropriate technology becomes an important
component of UN programmes on science and technology.

The CSTD itself should play a more active role in
future. Its agenda should be of a practical instead of
a deliberative nature. It should also ensure that the
combined UN system-wide efforts in the application of
science and technology to development are much more
effective. Increasing attention has been paid to science
and technology activities in the past few years. The
magnitude and variety of these activities in the UN system is also on the increase. The performance of CSTD
and of the ACC Sub-Committee can therefore improve only
if a more strengthened secretariat is provided to them.
The Office of Science and Technology, their present
secretariat, needs more staff and finances to perform its
servicing functions effectively. Although discussions
within the ACC Sub-Committee on Science and Technology
have taken a fortunate turn in respect of joint planning,
many problems still remain. In discussions on joint planning of programmes and projects, it would be desirable to
seek participation of those developing countries which are
interested in hosting (pilot) projects and programmes on
appropriate technology. Moreover, potential donors should
also be invited to participate in the discussions of the ACC
Sub-Committee when external financing is sought for joint

programmes and projects. The donor participation is likely
to facilitate arbitration of financial allocations between
agencies. To enable this joint inter-agency action, the
establishment of a special ACC Sub-Committee on Appropriate
Technology would certainly help to ensure that UN agencies
focus their attention on this important issue. The effective functioning of this Sub-Committee can be assured by
restricting its membership to those UN agencies which are
willing to accept the rules of joint planning.

The restructuring of the UN system, currently underway, is
likely to affect the future of ACAST, CSTD and ACC. There
is some feeling that ACAST, in its present form, will disappear. We believe that redesigning of ACAST may be necessary in the light especially of the outcome of UNCSTD in
August 1979. Yet it would not be desirable to abolish this
advisory body which could play an important role. It is our
belief that the types of changes proposed above in respect
of CSTD, ACC and ACAST, are fully in line with the main
objective of the restructuring exercise, namely, to make
the UN system more effective.

So far we have discussed only the programme component of
UN Science and Technology activities. However, the project
component at the country level, which has received much less
attention, is also relevant here.

In fact, there seems to be even much less coordination among
the agencies at the field level. Fortunately, the UNDP
field structure in each country could exert greater pressure
to ensure coordination of field activities relating to
Science and Technology. Measures need to be taken to enable
UNDP to play this role. UNDP should increase its own contribution to appropriate technology projects by devoting the
bulk of its budget for global projects for this purpose.
Consideration should also be given to raising the size of
this budget. UNDP could also play a catalytic role in
sensitising the governments to appropriate technology
through its country programming. To conclude, we believe
UNDP can make a coordinated, meaningful and multi-disciplinary contribution to the development and transfer of technology for socio-economic development.

Chapter 5
INTERNATIONAL MECHANISMS FOR APPROPRIATE TECHNOLOGY

F. Stewart

INTRODUCTION[3]

It is now widely acknowledged that there is a need for
appropriate technology to form a significant part of the

---

[1]Writing this just two days after the death of
E.F. Schumacher, I would like to record the very
substantial inspiration he provided to all who work
in this field - without him indeed it is quite likely
that this and many similar papers would never have been
written: I dedicate this paper to his memory.

[2]Senior Research Officer, Institute of Commonwealth
Studies (Oxford) and a Fellow of Somerville College,
Oxford.

[3]While working on this paper I benefited greatly from
discussions with R. Steinberg and C. Weiss at the World
Bank; R. Muscat, B. Harland and J. Berna at UNDP;
J. da Costa of UNCSTD; A. Bhalla of ILO; E. Owen of the
Appropriate Technology International; I. Ritchie of CGIAR;
G. McRobie of ITDG; M.E. de Boot of VITA; and V. Rabinowitz
and J. Davenport of the National Academy of Science. I
also benefited from comments from participants at the
Expert Meeting on International Action for Appropriate
Technology held in Geneva, December 1977. Needless to
say none of them are responsible for my conclusions.

investment programmes of poor countries.[4] The use of
advanced country technology in LDCs has had some severe
distorting effects on both their patterns of production
and consumption. The resource requirements of advanced
country technology has involved high levels of investment
per man, concentrating LDCs' scarce investment resources
on a small proportion of the population with the consequence of un- and under-employment and low productivity
for the rest of the population. On the consumption side,
the transfer of advanced country technology has tended to
involve inappropriate high income products, neglecting
the sort of products required to meet the basic needs of
the majority of the population.[5]

While the need for appropriate technology is acknowledged,
there are many obstacles which impede its introduction on
any significant scale. These may be classified into
obstacles arising on the supply side and obstacles arising
on the demand side:[6] that is to say, for successful introduction, those who make investment decisions in LDCs must
want to choose appropriate techniques (i.e. demand it) and
such techniques must be available. This paper is chiefly
concerned with the international dimension of the supply
side, but it is useful to put this into context by first
briefly considering the overall demand and supply conditions
necessary to get a successful appropriate technology programme under way.

---

[4]Some of the international interest in appropriate technology is catalogued in Cooperative Programme of Action
on Appropriate Industrial Technology, Report by the
Executive Director of UNIDO (ID/B/188, 17 April 1977).

[5]For a much more detailed analysis of the effects of
advanced country technology in poor countries and a
discussion of the characteristics of appropriate technology, see F. Stewart, Technology and Underdevelopment,
Macmillan 1977, Chapters 3 and 4.

[6]See C. Cooper's contribution to the International
Economic Association Conference on Appropriate Technology,
Teheran, September 1976.

If those who make investment decisions in LDCs are to choose appropriate techniques and products, three conditions must be met. First of all, the structure of demand must be such that there is a market for appropriate products. Secondly, the structure of incentives - of labour and capital costs - must be such that the appropriate techniques are also the profitable techniques. Thirdly, decision-makers must have easy access to information about appropriate techniques - at least as easy as access to information about inappropriate techniques. The first two conditions impose requirements for government policy in LDCs, related to income distribution, market structure and product promotion, international trade policy, government expenditure and price/tax policy. There is little that can be achieved by international action on the supply side without national policies to restructure the choice of technique towards appropriate technology. But however appropriate the pattern of demand and the system of prices and incentives, entrepreneurs will not choose appropriate techniques if they do not know about them. Thus appropriate channels of communication are of vital importance to an appropriate technology policy.

This brings us to the supply side of appropriate technology. A critical obstacle to the introduction of appropriate techniques are weaknesses on the supply side in relation to development and to communication of appropriate technology. Historical neglect of research and development into appropriate technology has meant that advanced technology is often the only feasible and efficient technology available. Today it is estimated that only 4 per cent of world R and D takes place in developing countries (including China); this represents a sizeable increase over the previous estimates, when only 2 per cent of the world's R and D (excluding China) was in LDCs.[7] Moreover, much (probably most) of the research in developing countries is unrelated to appropriate technology, being basic research or applied research

---

[7] Estimates prepared by J. Annerstedt for the OECD. The figure of 4 per cent for 1973 includes China. Excluding China to make the figure comparable with the OECD 1964 estimates, LDCs accounted for 3.6 per cent of world R and D in 1973.

designed to duplicate and compete with developed country
research. Thus an important part of an appropriate technology policy consists in the promotion of appropriate
R and D both within LDCs and elsewhere.

In recent years, there has been growing interest in the
development and identification of appropriate technology.
For example, the ILO publication entitled Technologies for
Basic Needs identified over twenty institutions concerned
with appropriate technology for rural and small-scale
activities.[8] Economists making empirical studies of the
choice of techniques have identified appropriate techniques
in a large number of industries.[9] Thus, although there is
undoubtedly the need for a major research effort, it appears
that in some fields appropriate techniques are becoming
available. But there remains a very serious communication
problem. If they are to be introduced, entrepreneurs in
LDCs must have easy access to the techniques - both to
knowledge about the techniques and to suppliers of the
necessary inputs, especially machinery. The communication
problem has both a national and an international dimension.
At an international level virtually all channels of communication are concerned with the transfer of advanced country
technology, not with appropriate technology. By far the
greatest proportion of international technology transfer
takes place commercially, via the multinational corporations,
technology licensing contracts, machinery suppliers and
engineering consultants. For the most part, these sources
of communication and transfer are concerned with modern
technology from advanced countries. Governments of developed

---

[8]Hans Singer, Technologies for Basic Needs, ILO Geneva, 1977.

[9]See G.V. Jenkins, Non-Agricultural Choice of Technique: An
Annotated Bibliography of Empirical Studies, Oxford Institute
of Commonwealth Studies, 1975; A.S. Bhalla (ed.), Technology
and Employment in Industry, ILO, Geneva, 1975; "The Choice
of Technology in Developing Countries", World Development,
September/October, 1977; N.S. McBain, The Choice of Techniques
in Developing Country Footwear Manufacture, London, HMSO,
1977; and M.M. Huq and H. Aragaw, Leather Manufacturing in
Developing Countries: A Study in Technical Choice, David
Livingstone Institute, University of Strathclyde (forthcoming).

countries - in their aid and training programmes - provide
another source of international technology transfer. In
the past, these governments have been accused of transferring inappropriate techniques, but recently a number of
countries have made serious attempts to transfer appropriate techniques. Notable among these is the US Appropriate Technology Program.[10]

Individual governments and international institutions are
coming to recognise the need to provide better communication
channels for appropriate technology and to do something
about it within the areas they control directly.[11] But
there is no systematic mechanism whereby information about
appropriate technology may be collected and transferred
internationally. Consequently, individual groups on
appropriate technology find themselves working very much
in isolation and people in LDCs who wish to gain information
on appropriate technology have little idea where to go. In
the area of appropriate technology there is a peculiar need
for systematic international channels of communication.
This is partly because the sources of appropriate techniques
are so diverse both in nature and in location. They include,
for example, old techniques from the now developed countries:
as has often been pointed out the science and technology
museums in the developed countries provide a very rich
potential source of appropriate technology; many traditional
techniques from one part of the world may be appropriate

---

[10]USAID, Proposal for a Program on Appropriate Technology,
US Government Printing Office, Washington DC, July 1976.

[11]For example, Resolution 2 on appropriate technology
adopted by the Second General Conference of UNIDO (Lima,
March, 1975); The World Bank, Appropriate Technology in
World Bank Activities, Washington, July 1976; USAID Program
on Appropriate Technology, Washington, July 1976; Appro-
priate Technology: Report by the Ministry of Overseas
Development Working Party, Overseas Development Paper No. 8,
London, HMSO, May 1977; Progress Report on the Application
of Intermediate Technology in the Inter-American Develop-
ment Bank, May 1977; and Appropriate Technology and its
Application in the Activities of the Asian Development Bank,
Occasional Paper No. 7, April 1977.

elsewhere; improvements in traditional techniques and
newly-designed appropriate techniques provide a third
source, emanating from local entrepreneurs, research institutes, universities and appropriate technology groups
both in LDCs and DCs. Without effective channels of
communication, many of these techniques will be applied
only in very limited areas, while elsewhere - sometimes
even in a neighbouring country - people are facing similar
problems to which similar solutions would be appropriate.
Because of the nature of the main sources of appropriate
technology few commercial groups are concerned with communicating and transferring the technology. This is the
second reason why there is a need for non-commercial
international action. Thirdly, the demand for appropriate
technology is very often absent, latent or ill-defined.
Hence a bigger effort on the communication side is needed
to overcome deficient demand. Fourthly, the sort of decision-makers, to whom appropriate technology is especially
suited are relatively small-scale, often located in the
rural areas and operating their enterprises on a family/
self-employment basis. They are in a particularly weak
position to seek out information for themselves.

There is as suggested above, a national as well as an
international communication problem. International channels
of communication are needed because the supply of and knowledge about appropriate technology arises in many different
countries, while the need for appropriate technology is
similarly internationally dispersed. But there is also a
communication problem within nations. As just suggested,
appropriate technology is to a large extent designed for
the small scale and rural, which presents a major internal
communication/dissemination problem. A national communication effort is needed for this. Something like a three- or
four-tiered institutional structure may be needed. Some
international institutional mechanism is required to collect
and transmit information from various national and international sources of appropriate technology to the national
or regional institutions. But the regional or national
institutions must then transmit the information to the
users; this is likely to require institutions below the
national level - let us call them local institutions -
located in various sub-regions of each country to provide
close contacts with users.

It should be emphasised that information must go both ways.

Problems and responses of users have to be communicated to those collecting information about appropriate technology and to those developing appropriate techniques, so as to ensure that the right sort of techniques are developed and transferred. Any sort of rigid institutional structure would be totally inappropriate. Flexibility of structure and of informational network is required, so that local institutions may communicate directly with the international mechanism, or with national or local institutions in other countries, to produce a rapid and efficient communication system.

The aim of this paper is to consider possible new international mechanisms that might be developed to meet the need for improved channels of communication for appropriate technology and to promote the development of appropriate technology. The national counterpart to these channels, briefly sketched above, is considered in detail elsewhere,[12] and will only be touched on below in so far as it affects the working of the international mechanisms. The need to accompany any international mechanism by national policies both in relation to the structure of demand for technology, and in relation to communications within the nation, must be underlined. Failure to change demand conditions so as to favour appropriate technology, or to introduce appropriate national and local communications, may thwart any international mechanism, rendering it ineffective.

The need for new international mechanisms for the promotion of appropriate technology has been recognised in a number of places and a number of proposals have been put forward. Moreover, in specific fields - particularly related to agriculture and some aspects of rural development - various international mechanisms (in particular Consultative Groups) have been in operation for several years, generating valuable experience which may provide a model in developing mechanisms to cover the much broader field. The workings of some consultative groups has therefore been examined. We also analyse and contrast the objectives and functions that might be met by international mechanisms for appropriate

---

[12]See Chapter 3.

technology. Finally, we present some conclusions about
requirements for new international mechanisms for the
promotion of appropriate technology.

SOME PROPOSALS FOR INTERNATIONAL MECHANISMS[13]

Inter-agency Network for Exchange
of Technological Information

The United Nations Inter-Agency Task Force established
under paragraph 6 of resolution 3507 (XXX) of the General
Assembly, proposed a network for the exchange of technological information, and this was endorsed by the General
Assembly. The network was to consist of "a large number
of individual nodes and links between them; the nodes should
be sources of technological information at the national, regional and international level in the private or public
sector; and the links would make possible the exchanging of
knowledge of the operations of all participants, compati-
bility among those operations, practical working arrangements and common aims. Each of the nodes would be an active
participant within the network, committed to cooperating
with any and all other nodes in seeking out ways of
accelerating and increasing the flow of technological
information" (Para. 5).[14]

---

[13]Numerous other mechanisms not described in this section
are directly or indirectly involved with transmitting information on technology. These include, for example,
UNISIST (UNESCO's Universal System for Information on
Science and Technology); AGRIS (The FAO International Information System for the Agricultural Sciences and Technology); the Control Data Corporation in the USA. The first
two are described in A.S. Bhalla and F. Stewart, "International Action for Appropriate Technology" in Tripartite
World Conference on Employment, Income Distribution, and
Social Progress and the International Division of Labour,
Background Papers, Vol. II, ILO, Geneva, 1976.

[14]Report of the Secretary-General to the United Nations Economic and Social Council, Institutional Arrangements in the
Field of the Transfer of Technology, Establishment of a Network for the Exchange of Technological Information, E/6002,
2nd June 1977.

"Thus the network should be viewed essentially as an international programme of cooperative action designed to support the activities of national and regional information services. Furthermore, it should not replace any of the existing or planned sectoral, national, regional or international networks. Rather it should be viewed as a means of strengthening these and providing, through shared knowledge, of the operations of all participating organisations, the means for their intercommunication" (Para. 6).

The proposed network would not "depend on any specific physical interlinking ... the participating organisations would themselves decide on the most suitable means of communications" (Para. 7).

"For the continuing day-to-day operation of the network, it would be desirable to have a series of well-designed comprehensive use-oriented sectoral, regional and international directories. These would be designed in such a way as to help participating organisations to identify the best sources of technological information and to find how to gain access to them and at what cost if any. The operation of the network and its effectiveness would be further enhanced by the development of training manuals, guides, etc. and by the periodic convening of seminars and workshops by the participating organisations" (Para. 8).

The Secretary-General and the Inter-Agency Task Force were asked to ascertain the current availability of information capabilities at the national, regional and international levels in terms of information sources, means of access to information and information related services, in order to "identify any deficiencies that might inhibit the setting up of the network". They made further recommendations in a report presented to the General Assembly at its 32nd Session which is complementary to the previous reports of the Task Force.

This report[15] concludes that the original concept of the

---

[15] UN Draft Report of the Secretary-General on the Establishment of a Network for the Exchange of Technological Information. Document "CO-ORDINATION/R.1236/Add. 1" of 23 September 1977.

network remains valid and that its establishment is
feasible.

"The effectiveness of such a network depends on a number
of actions which have to be taken on a continuous basis
for:

(a) The identification by countries and regions of the
    type of information which is relevant, its appropriate-
    ness to their specific needs and overall development
    objectives, and applicable at the various levels of
    end-use, namely policy and overall decision-making,
    use by scientific and other institutions and by in-
    dustries and ultimately assimilation by the less
    privileged populations in urban and rural areas;

(b) The adaptation by national and regional institutions
    of their means of dissemination of information ... to
    meet the needs at the various levels of end-use
    mentioned above" (Para. 52).

It is undoubtedly due to the complexity of the matter that
everything remains vague. It appears however reasonable to
conclude that it is not intended to establish a network of
a physical nature but rather a comprehensive programme
designed to foster greater cooperation between existing information generators, transmitters and receivers. The explicit identification of urban and rural poor as a main
category of end-users of technological information is worth
noting in this respect: "One of the reasons why basic needs
have not been met so far lies in the use of inappropriate
technologies. One of the means to meet the needs of this
end-user category is to be found in adequate information
delivery, starting from the premise that information should
be adapted to the requirements, the capabilities and the
means of the user, rather than vice-versa" (Para. 16).

Industrial and Technological Information Bank (INTIB)[16]

The proposed bank would form part of the international net-

---

[16]UNIDO, Establishment of an Industrial and Technological
Information Bank, Report by the Executive Director, ID/B/183,
April 1977.

work for the exchange of technological information, just
described. The proposal for the Bank was a response to
the recommendation of the Lima Declaration and Plan of
Action on Industrial Development and Cooperation, that
"Appropriate measures, including consideration of the
establishment of an industrial and technological information bank, should be taken to make available a greater
flow to the developing countries of information permitting
the proper selection of advanced technologies".

The Bank would be "concerned primarily with the selective
processing of technological information, thereby providing
the developing countries with a basis for making decisions"
(Para. 16); it would not only collect information but also
assess it. "The provision of information might in certain
instances have to be followed up by further consultations
and advice in the countries concerned with the aim of
assisting the recipient in assessing and applying the information" (Para. 18). It is intended therefore that the
Bank will not simply be an archive of information but will
provide access to information, its analysis and synthesis,
with the capability of giving on-site technical advice to
assist the recipient in the effective application of information (Para. 19). The Bank would draw on technological
information from within UNIDO and outside - e.g. information
from reports of technical assistance projects, consultants'
reports, reports of staff missions and of expert group
meetings, UNIDO research studies and information obtained
through the Industrial Inquiry Service, journals, library
and documentation. The Bank would not duplicate external
sources of information, but would establish a memory/
retrieval system for collecting relevant information when
required.

A Pilot project has been initiated, selecting four areas -
iron and steel, fertilisers, agro-industries, agricultural
machinery and implements. This stage is intended to last
from July 1977 to December 1978. Information will be
supplied in anticipation of demand when directly solicited
by individual requests. The target customers include
ministries of industry, planning and industrial development,
multipurpose technological institutions, transfer of technology centres and registries. "Our intention is to serve
all those who can be identified as having genuine technology selection responsibilities and problems, whether

in an advisory or decision-making capacity".[17]

While it is clear that the Bank would provide information
on appropriate technology - indeed its role is mentioned
in the documents related to the "Cooperative Programme of
Action on Appropriate Industrial Technology" (ID/B/188) -
its function is to provide information on <u>all</u> types of
technology and it has no special role in relation to the
dissemination of information about appropriate technology.

Technology Referral Service (TRS)

This World Bank proposal was put forward in the belief that
"International assistance and coordination can perhaps be
most useful in broadening the scope and quality of information available to developing countries for rational choice
of technologies ... The Bank would favour a system that
is initially modest, well-defined, pragmatic and fast action.
It would be a major boon to developing countries to make
readily available a mechanism to review the shelf of existing technological processes and associated hardware, and
present clearly and quickly the relevant elements permitting
informed choice of the technological alternative. Subsequent feedback to monitor the choice to see whether the
investment project is successful, beneficial to the in-
vestor and to the country should be provided for. The network would then progressively improve the knowledge base on
appropriate technology and help identify priority areas for
R and D".[18]

The TRS is designed to provide such a network; the TRS
proposal lays particular emphasis on the need to improve
the quality of technological choice, the need for any
information system to be closely linked with decision-makers,
and the need to follow up studies to see that the system is

---

[17]Letter to colleagues within the UN from R.T. de Mautort,
Industrial Information Section, July 1977, UNIDO.

[18]World Bank, <u>Technology Transfer and Appropriate Technology</u>,
<u>The World Bank Role</u>, 1977, p. 5 (mimeo).

working effectively.

Initially, the information provided would be channeled only
to private business or public enterprises needing help to
make decisions; and only to those where an investment decision is imminent. TRS should initially serve both large
and small enterprises "although the character of the service
may differ markedly". It is proposed that to start with the
service should deal only with industry, which it is felt is
particularly poorly serviced.

The TRS would operate with a clearing house, to which LDC
enquiries would be funneled through "screening institutions".
The screening institutions which should have a direct
interest in "appropriate" technology, might consist of a
financial intermediary (e.g. regional development banks,
and national finance corporations) or a national office for
the transfer of technology. The screening institutions
would be responsible for seeing that the enquiry is fully
specified, and that the TRS replies promptly and usefully.

The clearing house would receive enquiries, refer them to
technical sources, evaluate the procedure and recommend
improvements: it would have "the prime responsibility for
identifying, enlisting, refining, evaluating and where
necessary eliminating technical sources". Particular
emphasis is placed on the need for top professional talent
in the clearing house. It is not proposed that the clearing
house functions should be carried out in the World Bank, but
that some other existing institution should be used.

The World Bank proposal emphasises that the TRS could not
act as a complete substitute for on-the-ground technical
assistance, and the best approach would be to combine the
TRS with technical assistance. Emphasis is placed on follow-
up procedures to assess the value of the advice. "The
service, at least initially, would remain neutral as to
what constitutes "appropriate" technology for a particular
request. The clearing house role is to broaden the range
of information and the clarity and focus of that information.

The proposal contains fairly well worked out ideas for the
nature of the proposed processes of enquiry, response, monitoring formats, data organisation and storage, retrieval and
communications technology. It is our understanding that
this proposal is currently being reformulated as a research-

cum-pilot project in one country, to test the feasibility
of the TRS in a national context.

Socially Appropriate Technology Information System (SATIS)

This is a proposed networking system for the appropriate
technology groups: it was initiated by the Groupe de Re-
cherches sur les Techniques Rurales (GRET). The aim is to
systematise information storage/retrieval by the appropriate technology groups, so that on receiving enquiries
each group could call on the work of other groups. As the
proposal stands at present, it appears that SATIS would
involve classifying all technologies, recording details
of each on fiches, which would then be available to all
affiliated organisations. The Intermediate Technology
Development Group (ITDG) magazine, Appropriate Technology,
and TRANET (see below) already perform some networking
function among appropriate technology groups.

Transnational Network for Appropriate Technologies (TRANET)

This is a voluntary organisation whose main purposes include
promoting bilateral exchanges between appropriate technology
centres throughout the world and promoting the understanding
of, development and use of appropriate technology. "Towards
these purposes TRANET may operate a clearing house, publish
a regular newsletter, hold seminars..." (TRANET-BY-LAWS).

World Plan of Action (WPA) Fund[19]

In the UN World Plan of Action a special World Plan of
Action Fund was proposed to be established within the
framework of UNDP. The purpose of such a fund was to enable
UNDP to act as a catalyst in implementing research projects
which were given high priority in the WPA, in order to build
an indigenous science and technology capacity in developing
countries. This fund was supposed to absorb 50 per cent of

---

[19]UN, World Plan of Action for the Application of Science
and Technology to Development, New York, 1971, pp. 39-41.

the additional contributions to UNDP projected between 1970
and 1975, to amount to 25 per cent of the total resources
of UNDP (in addition to any current UNDP resources allocated to relevant projects under the country programming
and global project provisions). Thus the fund was not
proposed as a new special fund. In order to ensure flexibility and to take account of changes in research priorities the Governing Council of UNDP was to take the responsibility for changing the objectives of the fund.

Although the fund was never created, this proposal is relevant since its aim was to provide financial assistance
for research activities at "the national or regional levels,
for which such support was not otherwise available".

Sir Austin Robinson's Proposals for UNDP[20]

Robinson's report recognised the dual need for up-to-date
and comparable information on alternative technologies,
and for R and D to modernise traditional techniques and
devise new techniques appropriate to developing countries.
He proposed that a relatively small number of industries
should be selected (say eight to ten) to avoid dissipating
energies over a large field. In each of these areas international research institutes should be set up (analogous to
the seed institutes), each devoted to R and D in a specific
industry or industrial research centre, which had been
identified as "outstanding/excellent". "Each such international institute would be focused on the basic problems of
recording, analysing and developing technologies appropriate to the developing world in that one specific-area
industry, with the twin duties of being a complete source
of technical and economic information and experience regarding all existing practices and a centre for the development of the basic techniques". The institute's workshops would have examples of the principal techniques in
working order, and regular production, to identify problems
of production and maintenance. They would be linked with
institutes in the same field in advanced countries, which

---

[20]Future Tasks for UNDP - Report to the Administrator of
UNDP by Austin Robinson, (1976).

would undertake research for them when requested. Each individual LDC or group of LDCs would have an institute in
close contact with the world institutes to apply, adapt
and disseminate the results of the world institutes and to
identify problems of those engaged in the industries concerned.

A provisional, initial list of industries suggested was:
textiles, building and construction, farm machinery and
equipment, water pumping and distribution, footwear, food
processing and foundry work.

World Employment Conference Proposals for New International Mechanisms for Appropriate Technology[21]

One of the proposals at the World Employment Conference was
a new International Appropriate Technology Unit to provide
a means of coordinating and disseminating work on appropriate technology on a worldwide basis and to foster, encourage and disseminate new R and D to meet the basic needs
strategy. It was proposed that the Unit should be relatively
small (20-30) professional staff, and its aims would be focused on providing and improving technologies for the poorest
people in LDCs: it would collect and disseminate economic and
technical information about alternative techniques and products, monitor ongoing R and D, identify R and D gaps, find
means of getting the gaps filled and communicate the results
to users. The unit would not itself conduct R and D but get
R and D carried out in national institutes (in LDCs where-
ever possible) by identifying gaps and helping raise funds.
It would not duplicate information collection of other institutions but catalogue who knows what and where and provide a link between national (and other international) institutes where needed. National institutions would be put
directly into contact with each other, bypassing the central

---

[21]See A.S. Bhalla and F. Stewart "International Action for
Appropriate Technology", op.cit. and ILO, Employment Growth
and Basic Needs - A One-World Problem - Report of the
Director-General to the World Employment Conference, Geneva,
1976, Chapter 9 on "Technological Choice and Innovation for
Developing Countries".

unit wherever possible.

It was proposed that the Unit should select priority areas
starting with one or two, and covering at most say six at
any time: the priority areas would be selected in accordance
with various criteria including significance of industry,
geographical coverage, relevance to the poorest sections of
LDCs, and in the light of existing knowledge and suspected
gaps in knowledge. In the priority areas the Unit would
aim to be thorough and effective in collecting and disseminating information, in identifying R and D gaps, in
securing appropriate R and D and in communicating between
users and researchers. In non-priority areas, the Unit
should simply have the function of registering and communicating sources of information. This proposal was discussed
at the ILO World Employment Conference held in Geneva in
June 1976. At the Conference the Group of 77 and the
Workers Group endorsed the establishment of such a Unit.
Most industrialised countries however did not support the
proposal on the grounds that new institutions were un-
desirable.

A related proposal on the establishment of a Consultative
Group on Appropriate Technology was also discussed at the
World Employment Conference. The main functions of this
Group were to "suggest programmes of action taking into
account other programmes under way or being planned, and
to provide for their financing. The priority problem areas
could be identified from among candidates by small task
forces. Once a candidate problem area had been given a
priority by the Consultative Group, it could be entrusted
to an appropriate international or regional institute".[22]
The establishment of the Group was earlier recommended by
the ILO Technical Meeting on Adaptation of Technology
(organised in collaboration with UNIDO and ESCAP with UNDP
financial support).[23]

---

[22]See ILO, Employment, Growth and Basic Needs, op.cit.,
Chapter 9, pp. 150-152.

[23]See ILO, Policies and Programmes of Action to encourage
the use of Appropriate Technologies to Asian Conditions and
Priority Needs, Bangkok, 3-14 November 1975.

At the Conference the Group of 77 and the Workers Group
endorsed the establishment of a Consultative Group to be
especially directed to research on the choice of alternative uses of resources allowing a greater use of labour
per unit of investment ... [24] The Workers Group also
emphasised that the Group should be tripartite in character
including representatives not only of governments but also
of employers and workers. Most industrialised countries
did not support this proposal. In particular, the US de-
legate to the Conference felt that the proposal was premature since the institutional capacity in LDCs that was
so necessary for the success of such a mechanism was
lacking.

As the paper by Amulya Reddy (Chapter 3) shows, there is a
growing institutional capability in LDCs for undertaking
R and D on appropriate technology. It is often the wrong
priorities and lack of funds that prevent success of the
LDCs institutions in this area.

More recently, the Group of Non-Aligned Countries at their
Fifth Summit Conference held in Colombo in August 1976,
included an endorsement of the proposal on the Consultative
Group on Appropriate Technology, in their Action Programme
for Economic Cooperation.[25]

An International Centre for Appropriate Technology[26]

A. Khan, of the International Rice Research Institute, put
forward a proposal for the establishment of an international

----

[24]ILO, Declaration of Principles and Programme of Action
adopted by the World Employment Conference, (WEC/CW/E.1),
para. 62

[25]Action Programme for Economic Cooperation: Fifth Conference
of Heads of State of Governments of Non-Aligned Countries,
NAC/CONF.5/S/4, Colombo, 19 August 1976, p. 13.

[26]See International Center for Appropriate Technology,
proposal to USAID, 1974 (mimeo).

centre for appropriate technology to be located in an
LDC, whose task was the development of products and
techniques appropriate to very poor rural areas - particularly arid and semi-arid areas. The Centre would aim
to cover "movement of farm products, low cost alternative
energy sources, and simple equipment to enhance the
quality of rural living, such as for brick, block, rope
and mat-making, handicraft tools, simple farm vehicles,
equipment for rural road-building and maintenance, solar
collectors, windmills, waterwheels, rural home water
supplies and implements for insect and rodent control".
The proposal was developed in the light of the success
of the machinery development project at IRRI. It was
argued that "there is currently no international centre
for industrial development that could serve as a base
for development of appropriate technologies for the rural
poor in arid and semi-arid LDCs".

The Centre would conduct research, development, ex-
ploratory development and testing and industrial extension. It would need to be located where there was
some local capacity for machinery production. Once
prototypes were developed they would be sent to other
countries for evaluation and testing.

It was argued that "Owing to limitations of organisation,
personnel and other resources, industrial research organisations in developing countries have generally not
been able to mobilise the innovative inputs required for
new product development. The industrial research institutions are mostly engaged in providing routine
technical services and information to local industries
and research of limited commercial potential". Khan
therefore proposed a new international research centre.
The Khan proposal fits into the general framework proposed
by Austin Robinson, and could be viewed as one of his
proposed international institutes.

An Interim Global Project Towards an International
Council for Appropriate Industrial Technology[27]

This proposal is for a relatively small interim (or perhaps
pilot) project devoted to appropriate technology. The project would have a core staff of about 5 experts, an executive committee and an advisory committee which drew from
practitioner institutions. The major functions of the
project would be:

(a) To stimulate serious policy studies in individual
    developing countries looking into the scope and
    need for government action to facilitate the develop-
    ment and application of appropriate technology; the
    main focus of the studies would be on the determinants
    of technological decisions;

(b) To provide technical and financial support of a
    limited scope for on-going local operational
    programmes;

(c) To explore and promote "larger" project possibilities
    of international significance, calling for ad hoc
    funding - particularly related to new institution
    building.

This proposal thus differs substantially from earlier
proposals in that it is primarily related to social/economic
policy determinants, rather than to the collection, dissemination and development of information on technologies.

World Technological Development Authority

The report of the Club of Rome[28] on proposals to reshape

---

[27]This proposal by M. Usui, then at OECD Development Centre
(Paris) was presented at a UNDP meeting on Appropriate
Industrial Technology held in New York in April 1975. No
action was taken on it or on any other proposals.

[28]See J. Tinbergen (coordinator), Reshaping the International
Order; A Report of the Club of Rome, E.P. Dutton and Co. Inc.,
New York, 1976.

the international order proposed that a World Technological Development Authority should be established, to
be backed up by an International Bank for Technological
Development. These institutes would be concerned with
global technological issues, not simply those within the
UN system. The Technological Development Authority would
be a "planning, programming and training organisation which
would carry out feasibility studies, devise detailed programmes of research and development, arrange for their
implementation in cooperation with the Bank, by contact
with the most appropriate experimental institutions ...
supervise the programme of work in each case and act as
custodian of such industrial property as might accrue, on
behalf of the participating countries".

USAID Proposal for a Program in Appropriate Technology

The objectives of the US Program, which was incorporated
in a new section, 107, of the Foreign Assistance Act,
are:

"(1) To promote the development and dissemination of
technologies appropriate for developing countries,
particularly in the areas of agriculture and rural
development, small business enterprise and energy;

(2) To identify, design and adapt from existing designs,
appropriately scaled, labor-intensive technology,
and policies and institutions directly related to
their use;

(3) To formulate policies and techniques to facilitate
the organisation of new small businesses;

(4) To engage in field testing of intermediate technology;

(5) To establish and maintain an information center
for the collection and dissemination of information
on intermediate technology;

(6) To support expansion and coordination of developing

country efforts in this field".[29]

The programme included a $20 m. fund. An independent non-profit making institution - the Appropriate Technology International (ATI) was established to administer the fund. The programme areas were defined under five areas:

1. Communication and coordination;

2. National policies for appropriate technology;

3. Appropriate technology projects in LDCs;

4. Education of relevance to appropriate technology;

5. US Business - finding means of involving US business in appropriate technology.

The programme of ATI is not intended to involve establishing elaborate administrative structures, in any of the areas.[30] The intention rather is to contribute under each head by supporting the initiatives of others (e.g. supporting net-working among appropriate technology groups), by financing projects which contribute to the general aims (e.g. seminars or educational programmes related to communication and education), and by adopting an innovative and flexible approach to the subject - exploring new possibilities, learning by experience and developing new approaches.

The US Program is a national, not an international effort. This, obviously, makes the precise model inapplicable vis-à-vis international mechanisms. In my view, a serious disadvantage of the programme is its national, developed country organisation. The programme may reinforce the views of those in LDCs who lack commitment to appropriate technology - and who, indeed, argue that it is an attempt to sell them an inferior technology, and one which will keep them permanently under-developed. In this way it

---

[29] USAID, Proposal for a Program in Appropriate Technology, op. cit., p. 2.

[30] An overview of ATI's objectives and functions is described in AT. International: An Overview, Washington DC, March 1978 (mimeo).

could actually be counter-productive in terms of the overall
development and acceptance of appropriate technology.
Whether this is the net effect or not will be partially,
but only partially, dependent on how it is administered.
It will also depend in part on other stances the US takes
in relation to the North/South questions - stances which
have nothing to do with appropriate technology, or the
administration of the ATI. This is one reason why an
international mechanism is needed - and one that contains
a large LDC element.

Proposals of a Working Party of the UK Ministry of Overseas Development

The Report[31] concluded that "in view of the importance now
attached to intermediate technologies by the developing
countries, of the role aid donors can play, of the direct
relevance of the subject to our aid policy of doing more to
help the poor, particularly in the rural areas, and of the
possible advantage to British industry, the current modest
level of assistance to intermediate technologies within the
aid programme should be increased".

Policy recommendations included intensifying existing technical cooperation and capital aid activities in intermediate
technologies, encouraging the strengthening of information
gathering and dissemination, the establishment of links between British producers and overseas markets, encouraging
the testing, monitoring and evaluating of prototypes, and
considering providing finance to firms in Britain and developing countries for producing intermediate technology
products. Not less than £1/2 million per annum were to be
set aside for three years for this purpose. A clearly
defined departmental responsibility was to be created
within the Ministry of Overseas Development to cover all
aspects of intermediate technology; a proportion of the
funds were to be allocated to the ITDG to work in cooperation with the Ministry's Special Units. The Report also
recommended that the Commonwealth Development Corporation
should be asked to extend its intermediate technology
activities. Like the US Program, the UK proposals are
primarily national in focus.

---

[31] Appropriate Technology: Report by the Ministry of Overseas
Development Working Party. HMSO, London, 1977, op. cit.

Consultative Groups

The Consultative Group form of organisation may provide a
model for a new international mechanism for appropriate
technology. Around 30 Consultative Groups have been
established at various times. These groups are intended
to "achieve more effective use of development resources,
especially by coordinating flows of external assistance,
both with respect to objectives and policies, and also by
providing forums in which measures could be discussed for
improving the performance of developing countries and of
governments and organisations and giving them assistance".[32]
While most of the early groups were primarily concerned
with individual LDCs, recent groups have been devoted to
broad problem areas, that transcend national or regional
boundaries. For example, the Consultative Group on Food
Production and Investment in Developing Countries and the
Consultative Group on International Agricultural Research.

There is considerable variation in the form that Consultative
Groups have taken. In general each member of the Group re-
tains the right to independent action; actions taken in the
Group are entirely voluntary, and procedures followed by the
Group are highly informal. Membership of the Groups is
normally self-selected, with countries joining to benefit
from information exchange and coordinated action. In some
groups members are expected to express financial commitments
in support of selected projects sponsored by the Group. In
many, assistance remains purely bilateral. Institutional
arrangements tend to be minimal, with very simple and broadly defined terms of reference.

Consultative Group on International Agricultural
Research (CGIAR)

This Group - which has been directly concerned with questions
of agricultural technology - perhaps comes closest to providing a model for a possible appropriate technology group.

CGIAR was founded in 1971 following the success of the Ford
and Rockefeller Foundations in financing the international

---

[32]Discussion paper for Item (i) from Meeting of the Ad-Hoc
Group on Rural Potable Water Supply and Sanitation.

institutes which began the seed revolution. The Group was
sponsored by the World Bank, FAO and UNDP. The Group has
29 donor members, and five members to represent the five
developing country areas. It has a small secretariat of
seven, and calls on the expertise of a Technical Advisory
Committee consisting of 12 experts in the field. The Group
has been notably successful in raising finance. Total
financial support is now around $80 m. and growing at about
30% per year. The Group finances 11 International Research
Institutes/Centres devoted to agricultural research. Each
of the centres is autonomous, and has training as well as
research functions. Each centre has close links with the
country in which it is located; most centres are situated
near an agricultural university or research station, and
some carry on joint programmes of research. In addition,
the centres have international links with other research
centres in other countries. The centres build up library
and documentation services for international reference.
They also organise seminars and worshops and help finance
relevant projects in the countries in which they are located.

Special Programme for Research and Training in Tropical Diseases

This body is still at an early stage. It is a global programme of technical cooperation with and service to
governments, developed in response to a demand for coordinated research on the control of tropical diseases. It
has two objectives: the development of improved tools
needed to control tropical diseases, and the strengthening
of biomedical research capability in tropical countries.
The sponsoring agencies of the Programme are UNDP, WHO and
the World Bank. Like CGIAR, the Programme will have a
Scientific and Technical Advisory Committee.[33] Two major
differences between this programme and CGIAR are that the
participating countries (i.e. LDCs affected by the diseases)
will be of greater significance in the Joint Coordinating
Board; and the intention, it appears, is to finance research in existing institutions, or by particular groups

---

[33]Memorandum of Understanding on the Administrative and
Technical Structures of the Special Programme for Research
and Training in Tropical Diseases, TDR/CP/78.3.Rev.1,
Geneva, 1978 (mimeo).

of scientists, not to establish new international institutes.

Consultative Group on Food Production and Investment
in Developing Countries

The main functions of the Group are to encourage larger
external resource flows for food production to developing
countries, coordinate activities of various donors and
ensure more effective use of available resources. Membership is self-selected, including international institutions,
donor countries and recipient countries. It has a small
staff (of five).

The co-sponsors of this Group are the FAO, the World Bank
and UNDP. The Group has been chiefly concerned with looking
at world trends in food supply and consumption, and identifying required policies to increase food production and investment. There is some feeling that the Group has not
been sufficiently specific in its contributions, and that
it is not doing very much that other institutions could not
also do. It appears that this Group is now being phased out.

                    *   *   *

The Consultative Group model is thus an extremely flexible
one; its structure, organisation and membership is informal
and fluid. It can be, but is not always, a powerful agent
for raising and allocating funds, and providing for international coordination of activities in the chosen area.

One implicit difference between the various proposals -
which has considerable bearing on the relevant mechanisms -
is the definition of appropriate technology. Just as
everyone is against sin, so all favour appropriate technology;
and in an analogous way, interpretation of what is meant determines the relevant action.

At one extreme there is the view that the appropriate technology is the technology a country would choose, given as
wide as possible a range of choice to choose from. The
task of promoting appropriate technology then is that of
widening the available spectrum of technologies of all sorts:
the proposed Inter-agency Network for the Exchange of Technological Information and the UNIDO Industrial and Technological Bank are based on this principle, aiming to increase

access to information about all types of technologies in
order to extend the basis of choice. Adopting this approach
to appropriate technology tends to lead to prime emphasis on
information collection and dissemination, little emphasis on
R and D or social research and little selectivity in the
types of information collected and disseminated. At the
other extreme is the view that appropriate technology is
labour-intensive, small-scale, designed to meet the basic
needs and raise the productivity of the poorest people in
poor countries. This is the basis on which most of the
Appropriate Technology Groups work; it is broadly the definition adopted by Khan for his proposed institute and by the
proposals for the World Employment Conference. Adopting
this type of definition tends to lead to emphasis on a
particular type of information collection/dissemination, on
the need for R and D to develop new appropriate techniques,
and for social/economic research into the determinants of
choice of technique.

The World Bank takes what it described as a neutral view
of definition of appropriate technology. In the description
of the Technology Referral Service its initial function is
described as "to broaden the range of information and the
clarity and focus of that information".[34] However, in its
description of World Bank activities in relation to appropriate technology, it is apparent that appropriate technology is used operationally in relation to labour-intensive
small-scale technology which uses local resources and meets
basic needs. In defining appropriate technology in that
report, the World Bank identifies four dimensions of appropriateness: appropriateness to goal, appropriateness of
product, appropriateness of process and cultural and environmental appropriateness.

For the purpose of the discussion that follows, it is
necessary to be clear on what is meant here by appropriate
technology. Precise criteria are impossible to devise.[35]

---

[34]World Bank, Appropriate Technology in World Bank
Activities, op. cit.

[35]See F. Stewart, Technology and Underdevelopment, op. cit.
Chapter 4.

But it is not accepted here that promoting appropriate
technology is simply a matter of extending the range of
choice, but of extending it in a particular direction.
A summary definition was provided in the USAID report.[36]

- "In terms of available resources, appropriate technologies
  are intensive in the use of the abundant factor, labor,
  economical in the use of scarce factors, capital and
  highly trained personnel, and intensive in the use of
  domestically-produced inputs.

- In terms of small production units, appropriate technologies are small-scale but efficient, replicable in
  numerous units, readily operated, maintained and repaired,
  low cost and accessible to low income persons.

- In terms of the people who use or benefit from them,
  appropriate technologies seek to be compatible with
  local cultural and social environments."

This definition, broadly, summarises the direction in which
choice needs to be extended - though I would wish to make
two qualifications to it. First, appropriate technology is
a matter of appropriate products as much as appropriate
techniques - i.e. products which are appropriate for low
income consumers, make use of local resources, and fit the
local environment. Secondly, while the emphasis on small-
scale production units is correct, many innovations related
to large-scale production - innovations both in terms of
product characteristics and techniques - would increase the
appropriateness of the technology. It would be a mistake,
at this stage, to exclude appropriate technology for large
scale industry.

Defining the promotion of appropriate technology as being
the extension of the choice of technology in a particular
direction is of operational significance in terms of
mechanisms. It means that information collection and
dissemination need to be especially focussed at the appropriate technology end of the technology spectrum; information collection/dissemination of much of very recently

---

[36]USAID, Proposal for a Program on Appropriate Technology;
op. cit., pp. 11-12.

developed advanced country technology would <u>not</u> form part
of the promotion of appropriate technology. It also implies a similar selectivity and focussing of efforts in
relation to the promotion of R and D. Finally, it suggests
the need for complementary social and economic research
to identify the obstacles facing, and conditions conducive
to, the successful adoption of appropriate techniques.

The adoption of this type of definition is not intended to
suggest that collection of information about advanced
country technologies is a useless activity: obviously, it
may be very useful and some of these technologies may be
the best for LDCs in their particular circumstances, and
therefore in one sense the appropriate techniques. Nor is
the definition intended to be restrictive in the way, for
example, that perhaps some of the Appropriate Technology
Groups' use of the term may be. It is not intended to focus
solely on the very small-scale and rural techniques. Labour-
intensive large-scale techniques producing inappropriate
products for sale on the international markets may be more
appropriate than similar capital-intensive techniques, and
their promotion may be part of the promotion of appropriate
technology. What is intended by the definition is to focus on
a strategy of promoting appropriate technology in a particular direction, so as to give some content to the strategy,
and some guide to the requirements for mechanisms.

FUNCTIONS OF INTERNATIONAL MECHANISMS

The description of proposals and existing mechanisms makes
it clear that there are a number of functions that international mechanisms may fill in relation to the promotion
of appropriate technology. The proposals and mechanisms
differ in the functions for which they are intended. Some
of the functions are linked; others are not. It will be
helpful therefore to start by distinguishing the various
functions of international mechanisms, considering how far
they are necessary, how interrelated, and whether a single
or a number of mechanisms are required.

The main distinct functions are:

(a) <u>Information</u>

    1. Collection of information on technological alter-
       natives.

2. Dissemination of information on technological
    alternatives.

3. Networking.

(b) Research and Development

1. Identification of R and D needs.

2. Identifying and organising relevant R and D.

3. Funding R and D.

(c) Social/Economic Research

1. Analysing determinants of technological choice.

2. Following up success/failure of efforts under
    (a) and (b).

3. Analysing relevant government policies and
    institutional changes.

In general terms there are obviously strong links between
many of these functions: for example, R and D needs cannot
be identified without first knowing about what is currently
available. A comprehensive search for what is available also
sometimes involves some research. Then information collection and dissemination needs to be linked with R and D in
order to communicate the results of the R and D. Clearly,
information dissemination must be associated with information collection, or the latter would be useless. But despite the strong links, the functions are distinct and may
be performed by different institutions - as illustrated by
the many variations contained in the proposals described.
The discussion below considers the extent to which such institutional separation is possible and desirable, and how
the various proposals complement or duplicate each other.

There are gaps in the fulfilment of all the above functions
in relation to appropriate technology. While some work is
being done with respect to all the functions in various
parts of the world, communication between those doing the
work is weak probably leading to some duplication of work

on appropriate technology. Coverage, in terms of industrial
and geographical area, is sporadic, and the total effort involved is inadequate. Action therefore is certainly justified in relation to all types of function. However, not all
of this action need be international, and much of it, by its
nature, must be national.

Information

Collection and dissemination of information about the existing "shelf" of techniques are conceptually distinct functions;
different expertise and contacts are needed. Information
collection in a particular field is best done by experts
familiar with the field and requires contact with the various
research institutes and suppliers of technology throughout
the world. Information dissemination has an essential _local_,
not just national, component, requiring contact (directly or
indirectly) with those who make the investment decisions.
However, it is also necessary to have strong links between
those who collect and those who disseminate information. On
the one hand, and most obviously, the disseminators have to
have access to what has been collected. On the other,
efficient collection of information cannot be carried out in
complete isolation from those who are to use the information:
what information is useful, and how it should best be presented can only be decided in the light of knowledge about
the circumstances of the users. It is therefore necessary
to incorporate systematic links between information collection and dissemination, and for this some sort of international linking system is required. But the main thrust of
the dissemination effort has to be national; some national
institutions are needed to see that the dissemination is
properly carried out.[37] Information collection has an international element in that the information about technologies
is contained in different countries, but it can efficiently
be carried out within a national institution.

Information collection - if it is to be useful - is not
simply a matter of recording technical details of various
pieces of hardware. The whole process of production needs

---

[37]See Chapter 3.

to be described including software aspects, like skill and
managerial requirements, marketing and so on. There is
other essential information that is needed - for example,
the costs of the various types of machine, running costs,
likely life, maintenance and repair, and so on, which do not
appear in the normal technical manuals. Moreover, even such
apparently 'technical' information as rate of output of
different machines are related to the conditions under which
the machine is operated. Thus to be of help to a decision-
maker in an LDC, the information needs to be focussed on his
particular needs and problems, taking into account the social
and economic circumstances; and needs to contain far more
than simply a technical description of the various alternative machines. These requirements have implications for the
most efficient mechanisms for information collection/
dissemination.

At the collection end, they suggest that the semi-automatic
compiling of technical information - as in some proposed
data bank proposals are not likely to be of much use to
users. This sort of system is unlikely to give users information in a form that they can use. Users would be
likely to be served much better if they were put directly
into contact with institutions/individuals who have considerable experience in the relevant fields, and who may
then, through personal contact, present the information in
a helpful and relevant way. At the local dissemination end,
it would seem that individual technology users - those
making the investment decisions - need to be assisted in
formulating their enquiries and appreciating the replies,
by institutions/technical assistance experienced in dealing
with the sort of information presented.

Various individuals and institutions throughout the world
are already collecting and documenting much relevant information in relation to particular areas - partly in the course
of answering queries from LDCs, like VITA[38] and ITDG, partly
during the process of formulating appropriate development
projects, like the World Bank and some aid agencies, partly
during economic/social/scientific research. Centralising
all this information in a single data bank would be a mammoth

---

[38]Volunteers in Technical Assistance.

task, and one of limited value. It would require continuous
updating, would, for reasons already stated, probably be of
limited use to users, who would lose rather than gain by
receiving the information from the centralised source,
rather than from the individuals/institutions responsible
for its collection in the first place, who should be able
to formulate the information in more relevant form.

The conclusions for international mechanisms are that networking and linking users and sources of information is the
appropriate function for an international mechanism, rather
than the compilation of technical data. Networking involves
establishing a directory of sources of technological knowledge and linking users with the relevant source. Users may
be able to address enquiries directly to the central unit of
the network, which then hands the enquiries on to the relevant sources, or supplies the users with relevant names and
addresses; or the users' enquiries may be filtered through
screening institutions, as in the World Bank proposed Technology Referral System. Any network is only as good as its
individual components, at both ends - information source and
user application. In the field of appropriate technology
both ends are weak. Collection of information on the existing shelf of techniques has been sporadic and unsystematic.
This is one reason why some proposals - e.g. the UNIDO
Industrial and Technological Information Bank - do include
provisions for the collection of information. Any centralised network system should - in the process of its work -
help identify gaps in the compilation of information -
indeed this should be one of its functions. It might then
take responsibility for seeing these gaps are filled -
raising funds for it, and organising work. However, the
identification of these gaps and filling them will also be
one aspect of a systematic R and D effort, as will be discussed below, so that responsibility for this type of gap
filling might in part be handled by the system designed to
promote R and D.

Research and Development

The dividing line between collection of information about
the existing shelf of techniques and research and development into new techniques is a bit arbitrary, possibly un-
justified. The use of the term 'shelf' is itself misleading, as if all the alternative techniques were laid out

on a shelf in the larder, waiting to be identified and
catalogued. In fact a systematic search for alternative
'known' techniques can be difficult and expensive - it
may involve searching among the science museums for old
methods of doing things, as well as a geographically extensive search for methods currently in use. As this sort of
search proceeds, the dividing line between what is 'known'
and what is not tends to disappear, as does therefore that
between cataloguing existing information and discovering
new. Moreover, small changes in 'known' techniques (e.g.
use of different materials) may make the techniques appropriate in particular circumstances; or synthesising processes
in a new way - as shown by the research at Strathclyde
University - may make a major difference to the appropriateness of the whole productive process, e.g. in terms of
employment implications. But even minor adaptations or new
syntheses are innovations requiring engineering knowledge
and some field testing before they are proven. In many
cases, then, efficient search, or collection of information
also involves a research and development element. And, of
course, efficient R and D requires a preceding search/collec-
tion effort in order to identify requirements and gaps and
avoid duplication.

This discussion has relevance for institutional forms: it
suggests that search and research should be kept together
where possible. It supports the conclusions of the previous
discussion that a technology referral/networking service is
likely to be of greater use than a data bank, because the
former can link users directly with researchers, who will
have up-to-date information, and will be able to appreciate
how small technical innovations may improve or make possible
certain approaches; at the same time, the receipt of this
sort of enquiry will direct the attention of researchers to
current needs and problems. Some technologies are relatively
routinised and for these cataloguing of information may be
disassociated with research activity. A flexible networking/
referral system would allow for this.

Organising R and D for appropriate technology is probably one
of the most important aspects of an appropriate technology
policy, and one to which least attention has been paid. As
suggested earlier, there are three aspects to its identifying
R and D needs, getting particular organisations/individuals
to carry out the research, and funding the research. The
identification of R and D involves, as already stated, a

search for existing techniques. It also requires knowledge
of the social/economic conditions of the likely users of
the technology, as well as of their technical capacity,
their resources, managerial capacity, infrastructural
services and so on. Hence any successful research must
have strong links with likely users and a strong LDC
element. There has been some discussion about the relative
merits of carrying out the necessary R and D in developed
countries as against LDCs, and in international as against
national institutes. There are very strong arguments in
favour of R and D for appropriate technology being carried
out in LDCs: not only does this vastly ease the communications problem between users and researchers, but it also
means that the important learning effects of doing the R
and D are gained within the LDCs. The only case for carrying
out the research in the developed countries is that their
research capacity may be stronger. However, many LDCs have
large numbers of very good scientists and engineers. I would
conclude that appropriate R and D should take place in the
LDCs except in very exceptional circumstances. Of course,
international institutes may be located in LDCs - as for
example IRRI and Khan's proposed institute. The advantages
of international as against national institutes are that
they may be easier to fund internationally (along the lines
of the international agricultural institutes), and that
international use and communication of results may be easier.
Disadvantages are that there may be less national commitment
to the institutes, and that they are likely to be organised
on the lines of international organisations, with high
salaries, international bureaucracy and so on. An associated
question is whether it would be better to identify suitable
existing institutes (national) and get them to reorient their
work towards appropriate R and D, rather than set up new institutes. My feeling is that transforming national institutes would be of much greater long-run significance for
development, than creating new institutes, particularly new
international institutes. The drive for technological independence requires that the R and D should be done in and
by national institutes, rather than international. Currently,
the national institutes absorb large quantities of well-
trained scientific manpower. Yet it is widely believed that
much of the work is of little net benefit, consisting of
duplication of advanced country research, rather than independent appropriate research. If these institutions could
redirect their efforts towards appropriate technology, this
might make a major research contribution - far greater than

if the appropriate research were confined to a few international institutes. (It is estimated that 12% of the
world's scientific manpower is located in LDCs). Currently,
financial rewards and international reputations tend to go
to those scientists pursuing 'pure' science; and in the
applied field to those who produce techniques of use to the
large-scale 'inappropriate' technology sector. An essential
aspect of a policy to promote appropriate technology is to
change this.

Lack of major sources of funds for appropriate R and D has
been an obstacle. Despite much verbiage and the large
number of appropriate technology groups, total expenditure
on R and D is very small. Whoever does the research, funds
will be required. Currently, there is considerable potential
interest among donors - see for example the US Program, the
UK's recent endowment of ITDG, the Dutch government's concern
with the subject, and the documents indicating the interest
of the World Bank, ILO, UNIDO, UNDP, OECD and so on. Some
international mechanism is required to tap this potential
interest - (a) to survey the field fairly systematically to
identify potentially useful areas of research and individuals and institutions who would do the research; (b) to get
an international commitment to the use of funds so as to
avoid duplication of efforts by different donors; (c) to
incorporate a strong LDC element in decisions and organisation so as to avoid the 'colonial' taint that has touched
some of the work on appropriate technology; and to get
the essential national commitment to the development and
use of appropriate technology; (d) to make sure that the
results of the research are transmitted internationally;
(e) to organise follow-up studies to see how effective the
process was from identification of needs through research,
development, prototype testing, to introduction of the
techniques.

Social/Economic Research

There are strong reasons for believing that socio/economic
variables may be of as much significance as technological
ones in determining the choice of techniques.[39] Providing

---

[39]See F. Stewart, Technology and Underdevelopment, op.cit.

information about appropriate techniques (old ones or newly developed ones) may be necessary but not sufficient. A vital element of a successful appropriate technology policy may then be the identification of the other elements necessary for successful introduction of appropriate techniques.

PROPOSED AND EXISTING MECHANISMS: COMPARATIVE ANALYSIS

Information

The Inter-agency Network for the Exchange of Technological Information, the UNIDO Industrial and Technological Information Bank (INTIB), the World Bank's proposal for a Technology Referral Service (TRS), SATIS and the proposed International Appropriate Technology Unit are all intended to fulfil an information collection/dissemination function. Of existing mechanisms, VITA[40] and ITDG do a good deal of dissemination, with VITA's work being most systematically organised. The UNIDO enquiry service also does, but it is not directed at appropriate technology. The conclusions of the discussion in the previous section were that three functions were required of an international mechanism in relation to information collection/dissemination:(a) networking, rather than centralised cataloguing of information; (b) systematic dissemination including follow-up to see that the dissemination was effective; and (c) identification of information collection gaps, and getting these gaps filled.

The Network for the Exchange of Technological Information is still at a fairly early stage of development: it is possible that it could provide a framework into which other systems (e.g. TRS and SATIS) could fit. The Network may therefore provide some useful comprehensive framework, but as it stands it does not look as if it can do much for appropriate technology. The UNIDO/ITIB, as outlined, has

---

[40]VITA (the Volunteers in Technical Assistance) is a voluntary body situated in Washington which organises replies to technical enquiries from LDCs. They have 5,000 experts who are ready to answer the enquiries free of charge; they deal with between 1200 and 1500 enquiries a year.

a large element of information cataloguing, which, it has
been argued, is not the best way of collecting/disseminating
information. The UNIDO proposal, like the Network, has no
particular emphasis on appropriate technology. The most
promising networking proposal is that of the World Bank,
which contains carefully worked out proposals for networking,
for screening of enquiries, and for follow-up. Although the
Bank takes a 'neutral' view of appropriate technology, its
own work suggests that the system is likely to emphasise
appropriate technology, as understood here. It would seem
that the TRS offers the most well-worked out and promising
system for networking and dissemination, and should provide
the basis for a new international mechanism in this field.

There are three areas in which clarification and possibly
supplementation of the World Bank proposal are required.
One is to strengthen the appropriate technology side of the
service. As it stands it is possible that the service could
efficiently promote 'inappropriate' technology. It is proposed, to begin with, to service both small-scale and large-
scale enterprises but implicit is the proposal that it may
be desirable to phase out one or other of the two in the
next stage. It is essential, if the service is to promote
appropriate technology, that the small-scale side of the
operation is retained. Secondly, it would be helpful to
have built into the system some way of identifying informa-
tion/research gaps and some system to get them filled.
Further discussion is also necessary on the proposed
screening institutions. However, the TRS provides the basis
of a useful international network/enquiry service. SATIS is
a network for the appropriate technology groups, and could
fit, as a subsystem, into the TRS. One problem with SATIS
is that, as it stands, the proposal incorporates an informa-
tion-cataloguing as well as a networking function.

On the information collection/dissemination side, then,
while existing mechanisms are inadequate, proposed mechanisms
provide a satisfactory basis for the development of a new
international mechanism. However, while the TRS is likely
to provide a good formal network, effective information
communication also requires a vast array of informal activities - seminars, training sessions, journals, pilot projects,
consultancy and so on. No formal networking system can substitute for these. So far this sort of activity has been
organised by the Appropriate Technology Groups. Probably it
is desirable to keep much of them on an informal ad hoc

basis, but some activities could be organised/financed by
some international mechanism - e.g. the institution dealing
with the TRS - in the course of its other activities. The
USAID proposed Program includes informal dissemination
activities, as well as proposals to support more formal
networking activities.

Research and Development

The Network/information collection/dissemination proposals
discussed do not contain proposals for the promotion of
R and D. The proposals directly concerned with research
and development are Sir Austin Robinson's Report, the International Appropriate Technology Unit, the Consultative Group
on Appropriate Technology and the Khan proposal. These are
fairly different in content. The Khan proposal is for a
specific institute for research into appropriate techniques
in particular areas. While it obviously has much to be said
for it as a direct contribution to R and D, it has little to
do with the general problem of organising and funding appropriate R and D. Robinson's proposal is to establish a
number of International Institutes in chosen fields (of which
Khan's Institute could form one) - with links to national institutes - to collect and record information about existing
techniques and to research, develop and test new techniques.
The institutes would thus have the dual function of information collection and R and D, and could form part of any networking system adopted. The Bhalla/Stewart proposal is for
a small international unit, whose functions would be to
select priority areas, raise funds through the Consultative
Group mechanism, and commission research in those areas.
The research and development would be commissioned in existing national institutions in LDCs; the intention would be
for the international unit to act as a sort of catalyst
which, while directly financing some R and D, would encourage LDC institutions to redirect a large proportion of
their efforts in this direction. As argued above, there is
a strong case to be made for carrying out the required research and development in national institutions in LDCs.
The analogy with the seed revolution has inspired the idea
of international institutes. But the very much more hetero-
geneous field of appropriate technology may make the model
inappropriate. Moreover, some of the problems of the Green
Revolution - in particular the failure to take into account
social effects - may be due to the strong international

element in its development.

The Consultative Group on agriculture provides a further
possible model for international mechanism to promote
appropriate R and D. In agriculture in particular, CGIAR
has been notably succesful in fund-raising, and in chan-
nelling these funds into research. This Group has devoted
the funds to international institutes, but others (e.g. the
proposed group on tropical diseases) intend to use funds
raised to finance research in national institutes. It is
clear that irrespective of whether the research is eventual-
ly funded in new international, old national or new national
institutes, the first priority would be to get donors and
LDCs together to commit their funds and interest to appropriate R and D, and to decide on priority areas. For this
CGIAR provides an excellent model, being flexible and informal, lacking the bureaucratisation which often bedevils
international bodies. CGIAR has a small secretariat, and a
technical advisory committee consisting of distinguished
scientists in the field. CGIAR's membership is also fluid,
but until recently it has been dominated by the donors. A
similar group for appropriate technology would need a much
stronger LDC membership and commitment.

While the TRS provides a good starting point for information
networking, proposals for promoting appropriate R and D are
still at a much earlier stage, but elements of a fruitful
approach to the question are contained in the examples of
Consultative Groups. If a similar mechanism were established for appropriate technology, it would draw on the
other proposals in determining how to go about organising
R and D for appropriate technology.

Social/Economic Research

M. Usui's proposal is the only one directly related to the
socio-economic dimensions of appropriate technology. It,
or something similar, would play a vital role in comple-
menting any major effort on the technical side, providing
an opportunity to follow up and assess the effects of improved information and improved technologies. The Green
Revolution story indicates how important such research would
be.

CONCLUDING REMARKS

Recently there has been an upsurge of interest in appropriate technology; yet despite widespread approval of the
idea, action has been of almost minor significance. The
lack of action is not due to lack of finance; more to lack
of sufficient specific and well thought out ways of promoting appropriate technology. Much of the necessary action -
perhaps most - requires changes at the national level
within LDCs. But there is also an urgent need for new
international mechanisms to improve information collection
and dissemination and to promote appropriate R and D.
Despite the view voiced above, that it is difficult to
separate the two functions completely, it is possible and
desirable to separate the institutional form the new
mechanisms take, in order to make the task of manageable
proportions.

One need is for improved systematic networking. This
paper takes the view that the World Bank proposal for a
Technology Referral Service offers a promising model; and
should be supported. But the service should perhaps make
greater efforts to concentrate on appropriate technology
than appears in the proposed outline. It should also aim
at identifying information gaps and provide some way in
which they could be filled. Any institution responsible
for the networking should obviously have links with other
international mechanisms for appropriate technology. But
there is no need for them to form part of the same institution. The new networking mechanism would not be a substitute for the many informal information dissemination
activities which would continue to be of great importance.

Another major need is for the promotion of R and D into
appropriate technology. Very little systematic work has
been done here, and it seems likely that the potential
returns may be huge. It also seems that funds would be
available if suitable opportunities could be identified.
But we are not yet at the stage when it makes sense to set
up a vast array of new institutes. What is needed at this
stage is (a) raising funds; (b) identification of priority

areas; (c) organising finance of R and D in the priority
areas - selecting institutions/individuals most likely to
be productive; (d) organising prototype testing. A framework of action on each of these is spelled out in the
following Chapter.

Part III
Global Action

Chapter 6
A BLUEPRINT FOR ACTION

P. Henry, A. Reddy and F. Stewart

INTRODUCTION

Chapters 3, 4 and 5 have already established the scope for
much more concerted action on appropriate technology than
what obtains at present. Yet the existing institutional
mechanisms do not face up to this challenge. Since the
writing of the above chapters, we undertook extensive
field visits to a number of countries (e.g. India,
Bangladesh, Indonesia, Thailand, Philippines, Korea,
Kenya, Mexico, Japan, United Kingdom, Netherlands and
the United States), to United Nations organisations
(e.g. ILO, UNIDO, UNEP, UNCTAD, FAO, UNESCO, UNDP, UNICEF,
ESCAP and the World Bank), to appropriate technology groups
(e.g. TRANET, ITDG, VITA and TOOL) and to national science
and technology institutions in LDCs (e.g. Asian Institute
of Technology, El Colegio de Mexico, Technology Development Centre, Bandung, and Indian Institute of Science,
Bangalore).

The programme of action outlined here is generally supported
by the majority of those who were consulted during the field
visits. While some reservations were expressed on details
the proposal to establish a new international mechanism for
the promotion of appropriate technology (IMAT) received

---

[1]Secretary-General, Society for International Development,
Rome; Professor, Indian Institute of Science, Bangalore;
and Senior Research Officer, Institute of Commonwealth
Studies, Oxford.

overwhelming support.[2]

For the global action proposed here, appropriate technology is defined as technology which will raise the productivity and incomes of the poor in rural and urban areas, which generates productive employment, makes full use of local resources, and produces the types of goods and services needed to meet the minimum need of all the people.

The prime emphasis of such a definition is towards small-scale capital-saving technologies which are accessible to the poor, and which are essential for the creation of productive employment on a sufficient scale. But the definition also includes large-scale and relatively capital-intensive technologies which may also have a vital role to play in meeting development objectives.

In the light of the inappropriate characteristics of much technology emanating from advanced countries, we found widespread agreement that there is a need to promote appropriate technology designed to correspond to the conditions and needs of third world countries, in two main categories: (a) the adaptation of large-scale technologies to suit the requirements of developing countries; and (b) the development and dissemination of small-scale employment-generating technologies.

There are often serious deficiencies in the availability of appropriate technologies. This in part is due to lack of research and development (R and D) on appropriate technologies, and in part to weak dissemination of information about those appropriate technologies which are in existence. Current emphasis with respect both to R and D and to dissemination is overwhelmingly concerned with advanced country technologies. Hence there is a need for R and D and for improved dissemination of information regarding the hardware aspects of appropriate technology.

---

[2]This Chapter draws heavily on the feasibility study on IMAT prepared by the authors. See "A New International Mechanism for Appropriate Technology - Feasibility Study by a Team of Specialists". The Hague, October, 1978.

However, even when the hardware is available, it is well known that the dissemination and application of this hardware does not proceed as rapidly as required by development needs. Many social and economic changes are essential - for example, changes in access to resources and assets, in prices, in the lending policies of banks, in government regulations and expenditure policies, and so on. The socio-economic software aspects of appropriate technologies are not less, and sometimes far more, important than the hardware aspects; both aspects must be emphasised in the promotion of appropriate technologies.

National technological capability is a crucial component of the self-reliance of developing countries. Further, appropriate technologies are largely location-specific, resource-specific, and perhaps even culture-specific. For these reasons, the development and dissemination of appropriate technologies must be unequivocally based on national efforts in the developing countries.

These national efforts can, and do, emerge from established institutions of education, science and technology as well as from "voluntary groups" and "non-governmental organisations". Neither of these two mechanisms must be viewed as any less important than the other; each has its special strengths and weaknesses, and it is the utilisation of both which best serves a national thrust towards appropriate technologies. The effectiveness of this thrust depends upon the level of technological capability in the developing country.

Even when the level of national technological capability is high, the generation and dissemination of appropriate technology is limited by:

(a) an atmosphere in which little prestige is attached to work on appropriate technology in contrast to the active encouragement of work on advanced country technologies;

(b) an excessive bias towards R and D on advanced country technologies;

(c) the lack of identified priority areas in appropriate technology;

(d) the paucity of information relevant to priority areas;

(e) the weakness of the development phase of technology generation compared to the research phase;

(f) the absence (or deficiency) of appropriate technology delivery systems involving effective linkages between producers (institutions/groups/organisations) and users of appropriate technology;

(g) the lack of interaction and coordination between different appropriate technology efforts, for example, between the efforts of institutions and voluntary agencies or of national and local organisations in the same country, and between appropriate technology activities in different countries;

(h) the inadequacy of funds, equipment, etc. for critical activities such as pilot plants, field trials, demonstration projects, information exchange, etc.

When the technological capacity of a developing country is low, there are other serious limitations in addition to those listed above:

(a) the absence of appropriately endowed institutions with the potential for developing appropriate technology;

(b) the insufficiency of technical expertise;

(c) the absence of training facilities;

(d) the inadequacy of funds for R and D.

NEED FOR GLOBAL ACTION

It is these limitations on national appropriate technology efforts which define a perspective for global action. The essence of this perspective is that the principal, and perhaps even sole, objective of global action should be to strenghten national appropriate technology efforts where they are weak, and assist in initiating them where they do not exist. In other words, the primary purpose of this action is to help national efforts to overcome their limitations.

As noted in earlier chapters, many existing international
institutions are concerned with technology transfer from
the developed to the developing countries. At the same
time, some international institutions are directing efforts
towards the development and/or dissemination of appropriate
technologies. Despite this, the fact that the primary responsibility of these institutions does not lie in the field
of appropriate technology means that it forms only a peri-
pheral part of their activities. Neither are these international institutions concerned, to any significant extent,
with research and development directed towards the generation of appropriate technology. In short, no international
institution has the promotion of appropriate technology as
its sole objective.

Recently, national institutions for the promotion of appropriate technology have been established in some developed
countries - in the UK and USA - but, lacking an accepted
international character, they are unable to ensure the full
participation and commitments of developing countries.
Similarly, many small voluntary bodies in the developed
countries, e.g. VITA (USA), ITDG (UK), GRET (France) and
TOOL (Netherlands) are devoted to promoting and disseminating
appropriate technologies. Although these voluntary efforts
are useful, their magnitude is inevitably small in relation
to the massiveness of the task. In addition, they often lack
the necessary international and developing country component.

The foregoing discussion has led to four major conclusions:

(a) there is an imbalance in global work on technology
    with an overwhelming portion of this work not being
    primarily concerned with technologies appropriate to
    the needs of developing countries;

(b) whilst being the best guarantee and surest foundation
    for the successful development and dissemination of
    appropriate technologies, current national efforts
    within developing countries suffer from a number of
    limitations;

(c) there are severe deficiencies in the flow of information
    about appropriate technologies between nations leading
    to some duplication of efforts and providing an impedi-
    ment to the introduction of appropriate technologies
    where these exist;

(d) a suitable international mechanism for the promotion of appropriate technology is currently absent.

It follows that <u>there is a strong case for an international mechanism for appropriate technology</u>.

A NEW INTERNATIONAL MECHANISM FOR APPROPRIATE TECHNOLOGY

It is proposed that the new international mechanism should be a non-governmental institution. Although outside the main body of the United Nations system, it should be closely associated with it through a sponsorship arrangement. Thus, it would be a consultative group with a category "A" status which would give it the privilege of being consulted by the UN system in all matters related to appropriate technology. It would have the character of an association to provide the possibility of individuals and institutions becoming members of regional/sub-regional/national/sub-national chapters being formed.

Functions of the Mechanism

The basic objective of the new mechanism will be to help national appropriate technology efforts in overcoming their limitations.

The role of IMAT will therefore be <u>supportive</u> and <u>catalytic</u>: thus the aim of IMAT, in relation to each of its functions, would be to encourage and assist other institutions - national and international - to redirect their efforts effectively towards appropriate technology. The intention should be that IMAT would rarely take an exclusive role in any project; while it might initiate action it would always try to involve others at an early stage and aim to pass total responsibility to other institutions/individuals as soon as feasible. In playing this supportive and catalytic role, IMAT will have various functions which should possibly include:

(a) helping in the identification of priority areas for appropriate technology work;

(b) identifying institutions and groups which require

critical support for the successful development and
dissemination of appropriate technologies;

(c) providing suitable assistance by way of information,
funds, equipment, training, experts, etc. to these
institutions;

(d) assisting the passage from the research to the development phase in the generation of appropriate technologies,
e.g. through pilot plant trials, and from the technology
generation to the diffusion phase, e.g. through pilot
demonstration projects;

(e) strengthening appropriate technology delivery systems
by facilitating direct contacts between the producers
of appropriate technologies and the users of such
technologies;

(f) contributing to the generation of an atmosphere in which
the prestige of appropriate technology is enhanced;

(g) facilitating the exchange of experience among appropriate
technology institutions/groups in different countries,
sub-regions and regions;

(h) disseminating appropriate technology "success stories"
as well as insights into causes of failure of hardware
and/or software;

(i) assisting the creation of a new national or sub-
national institution when circumstances make such an
institution crucial to national appropriate technology
efforts;

(j) studying ways in which private efforts on generation
and transfer of technology might be made more appropriate both with respect to technology generated by
advanced countries, and with respect to technology
developed by local firms in developing countries;

(k) reviewing developments in the field of appropriate
technology including socio-economic aspects;

(l) carrying out all other activities, such as fund-raising
and monitoring of the effectiveness of its own efforts
to enable it to discharge the above functions.

The precise functions carried out by IMAT, and the balance of work as between different functions, is likely to vary over time as a result of experience gained and changes in conditions. Two functions - monitoring its own activities, and conducting socio-economic research into technological choice - will be of particular significance in helping IMAT to identify priority areas for its own efforts.

Operation

It is suggested that the operation of the new mechanism should be guided by the following principles:

(a) the principal aim of IMAT should be to support and catalyse appropriate technology efforts of institutions and groups in developing countries;

(b) IMAT should not attempt in-house R and D activities;

(c) IMAT should not create new international institutions for the development of appropriate technology hardware and software;

(d) where international action is required, IMAT should initiate an international dimension of existing national institutional activities, e.g. an institution in a particular country may be helped to become a training centre for the sub-region or region;

(e) instead of handling the whole spectrum of possible appropriate technologies, IMAT should concentrate on a few selected priority areas, e.g. appropriate agricultural processing technologies. The selection of these priority areas must reckon with current efforts of countries, institutions and groups active in appropriate technology, not only to respect their autonomy but also to benefit from their field experience. The criteria for selection of priority areas should clearly be consistent with the objectives of appropriate technology - viz. to raise the productivity and income of the poor and, to make use of local resources, to produce appropriate products to meet the needs of the poor and to generate employment; the choice of priority areas must also take account of the existing technological situation, selecting areas where there is a lack

of appropriate technologies, or where known techniques
are of low productivity. However, selection of priority
areas will vary according to which function IMAT is
pursuing; for example, information dissemination activities are best pursued where appropriate technologies
have already been identified in some part of the world;

(f) there should be considerable flexibility in approach
so that IMAT could interact with governmental and/or
non-governmental institutions or agencies in the developing countries and with national, sub-national or
local institutions. IMAT should not attempt to centralise appropriate technology efforts within a country.
Considerable flexibility is essential in view of the
competition and even conflict which sometimes exists
between different groups and institutions;

(g) many of the active appropriate technology groups and
institutions have achieved their present level of
efforts through a strong spirit of self-reliance. Since
such a self-reliant activity is a major objective of IMAT,
its linkages with these groups and institutions must be
designed to strengthen this self-reliance. This implies
that IMAT support to these institutions must in no way
lead to the imposition of priority areas, programmes,
experts, etc. In some ways, the collaboration between
the Development Technology Centre - Institution of
Technology, Bandung and the TOOL Foundation in the
Netherlands is a model which ensures the self-reliance
of the DTC-ITB whilst enabling TOOL to be of considerable
assistance;

(h) in facilitating the collaboration of different groups
and institutions, IMAT should aim at the establishment
of self-reliant networks of the type established in the
United Nations University Programme on Traditional
Technologies;

(i) IMAT should not aim to start operation in all developing
countries; the approach should be based primarily on
organic growth relying on institutions already involved
in appropriate technology activity or where it is believed such work can be promoted;

(j) since, however, such growth poles may not be well distributed over and within the various regions, it is

important to identify regions or sub-regions in which
appropriate technology activities are conspicuously
absent and to help to establish in these regions institutions dedicated to appropriate technology. Of
course, any new institutions which are created must
have characteristics which facilitate the development
and dissemination of appropriate technology, e.g.
proximity of target groups;

(k) insofar as the thrust of appropriate technology must
change with time and vary with location, IMAT should
not be preoccupied with constancy and uniformity of
approach. It should be dynamic and adaptable. Thus,
IMAT should be flexible in the definition and development of its own functions;

(l) it will not be the function of IMAT to coordinate the
appropriate technology activities of the UN and other
international agencies; but IMAT will seek to coordinate its efforts with these agencies when their work is
related to the same priority areas. In particular,
when these agencies, e.g. the United Nations University
or the Regional Commissions, already have programmes
which are in tune with IMAT's own objectives, IMAT
should support and work through such programmes provided
that they involve actual field activities at the grass
roots level.

IMAT may assist international institutions and national
developed countries' efforts to support appropriate technology by identifying suitable projects and activities.

Organisational Structure

Secretariat. The secretariat should be small since its
principal role is to support, catalyse and activate national
appropriate technology efforts in the developing countries.
In the first few years of operation, it may perhaps consist
of about 3-6 senior professionals with about one for each
priority area and the requisite support staff. To compen-
sate for this restriction in size, the secretariat may
convene ad hoc panels of experts to assist in special areas
and/or establish networks of groups active in those areas.

The members of the secretariat will spend a considerable
proportion of their time (e.g. half) away from the headquarters of IMAT, visiting national institutions, appropriate technology projects and so on. The staff members
should be technologists and social scientists. It is important that the secretariat consist of people who are
experienced, innovative and entrepreneurial.

The Director and senior staff should be appointed by the
Governing Body which in turn will be selected according
to procedures instituted by the founding members (see below).

Location. The location of IMAT's secretariat is bound to be
an important factor which will help determine its acceptability and performance.

It is clear from the thrust of the previous discussion that
IMAT must be located in a developing country; it should be
situated in a country which will welcome the institution,
and preferably, one that already has a serious commitment
to appropriate technology. It will be an advantage to IMAT
if the particular developing country has a rich experience
in appropriate technology so that IMAT's secretariat can
benefit from locally available intellectual inputs and
awareness of field conditions and problems. Of course, IMAT
should be located in a place with good transport and communication facilities.

It is not necessary that the location decided upon initially
should be a permanent one; the location for IMAT's initial
gestation or take-off period can be temporary. During the
course of consultations, many of the UN organisations seemed
to wish to host IMAT, locating IMAT either physically and/or
de jure within their auspices. However, the team received
very strong intimations that it would be preferable to have
IMAT outside the UN system, both formally and physically:
locating IMAT within any single existing UN organisation
would lead to counter-productive inter-agency rivalries.
This feeling was shared by many people consulted in the
developing and developed country research institutions and
government departments.

Notwithstanding the possible cost advantages of locating
IMAT in an existing international institution, such a move

would bring in its wake the disadvantages associated with
the image, procedures and influence of the host institution.
It appears therefore that IMAT should not be housed in any
UN agency, organ or organisation.

Whichever the initial location of IMAT, there may be a
tendency for it to concentrate its activities in the region
in which it is located. This tendency can be corrected by
IMAT having regional offices or perhaps using one of its
associated institutions to take on, with IMAT's assistance,
IMAT's supportive and catalytic functions for the region.

Governing Body. The Governing Body should consist of between
20 and 30 eminent persons who have made distinguished contributions to the field of appropriate technology. They should
not be representatives of either governments or agencies, but
would be selected as a result of recommendations from relevant institutions such as (a) donor governments/agencies,
(b) the governments of developing countries, (c) appropriate
technology institutions and (d) non-governmental bodies
active in appropriate technology. An attempt must be made
to ensure that the governing body achieves a reasonable geographical representation, and a balance in favour of the developing countries.

The Governing Body will select the Director of IMAT, and,
with the assistance of the Director, appoint senior staff.
The Governing Body and Director will define the constitution
of IMAT, in the light of discussions at the Founding
Conference. The main role of the Governing Body will be to
provide general directions for IMAT's work in its priorities
and programmes, and provide a critical review as the work of
the mechanism proceeds. The Governing Body will meet once
every year or two.

Executive Council. The Governing Body will provide overall
direction, but because of its size and the (relative) in-
frequency of meetings it may be desirable to have an executive council which meets more frequently to provide a more
regular overview of the work of IMAT. The Executive Council
would consist of about six persons, two-three from the
secretariat of IMAT and three-four from the Governing Body.
The Executive Council, which is ultimately responsible to
the Governing Body, should meet about every six months.

Level of Finance

To finance the secretariat, a minimum sum of US$ 0.5 - 1.0 million will be required. This sum will be used mainly for secretariat salaries and various administrative expenses including travel of secretariat staff to field projects and the organisation of governing body, executive council, expert panel and other meetings necessary for the secretariat to discharge its functions.

In addition, IMAT must have funds for the support and initiation of projects and programmes in the field. It is envisaged that the bulk of IMAT's funds will be devoted to these field activities. Administrative expenses should not exceed some fairly low fraction of the overall expenditure of IMAT. A figure of 10% was suggested. However, it is difficult to be precise about any such figure partly because what is desirable and feasible is likely to change over time as IMAT gains experience and the administrative machine is established. More significantly, because of the catalytic role of IMAT, IMAT may often initiate appropriate technology activities but not finance them, or only finance a very small proportion of the total activities. In such cases, IMAT's administrative costs may appear to be a high proportion of its total expenditure, although they are a low proportion of the total expenditure on appropriate technology generated by its work. Thus any target figure for the ratio of administration to other expenditure would be misleading and might even distort the activities of IMAT.

It is suggested that at a Founders' Conference, which will bring IMAT into being, a number of governments and agencies will contract to place at the disposal of IMAT a block grant of around US$ 10 million for an initial take-off period of three years. IMAT will prepare a three-year budget with annual components to be approved by a small Finance Committee. This Finance Committee which will include representatives of donor governments and agencies, is intended to assist IMAT's Governing Body in matters of finance. The approval of the three-year and annual budgets will be in terms of the directions, priorities and programmes.

Such an arrangement will ensure that IMAT's secretariat can concentrate during the take-off period upon substantive matters and programming work, rather than on public relations and fund-raising on a project-to-project basis.

It is envisaged that towards the end of this initial take-off period, IMAT's performance will elicit regular contributions from governments and agencies so that its activities can be expanded.

CONCLUSION

We recommend that the new international mechanism should be established as soon as possible. For this purpose, a special Founders' Conference should be organised, preferably to be financed by a number of participants including donor governments from developed and developing countries, and non-governmental appropriate technology bodies. The Founders' Conference should determine the functions, location and funding of the new mechanism. It should also establish and institute procedures for the appointment of the Governing Body of the mechanism.

The early establishment of the new mechanism would be very timely in relation to the forthcoming UN Conference on Science and Technology for Development to be held in August 1979, which should enable the new mechanism to elicit wider participation and support.